Dedication:

To teachers, students, and school leaders everywhere — this book is for you.

Foreword / Author's Note

The Role of AI in Modern Education

Education serves as the cornerstone of every society, yet those involved in the field—teachers, students, and administrators—often encounter significant and sometimes overwhelming challenges. These challenges range from restricted time and resources to the ongoing demands brought about by rapidly evolving technology.

This book emerged from a fundamental belief: artificial intelligence can act as a valuable partner in education. Rather than replacing the essential human elements of creativity and care, AI is intended to complement and support them.

The prompts provided within these pages are designed with clear intentions. For teachers, they aim to ease the daily workload, allowing more time for meaningful instruction. For students, the goal is to make learning in the classroom more engaging and accessible. For schools as a whole, these prompts offer ways to enhance communication and foster stronger connections within the educational community.

Each section of this book is carefully crafted to ensure that AI becomes an empowering tool—one that uplifts educators and students alike, rather than creating a reliance on technology. Integrating that, by integrating AI thoughtfully, all members of the educational system will find renewed support and inspiration in their important work.

— Mohammad Taib

Message for Teachers, Students & Administrators

Dear Teachers,

This playbook has been thoughtfully created to help you reclaim valuable time in your day-to-day work. Inside, you will find a collection of prompts that serve as templates, offering you innovative ways to inspire and engage your students. You are encouraged to adopt these prompts to fit the unique needs of your classroom environment.

By allowing AI to take care of repetitive and routine tasks, you free yourself to concentrate on what truly matters: teaching with heart, creativity, and dedication. Use this resource to streamline your workload, so you can devote more energy to fostering meaningful connections with your students and nurturing their love of learning.

Dear Students,

You are growing up in a world where learning never stops. The prompts in this playbook are not meant to be shortcuts; instead, they serve as boosters to help you on your educational journey. These prompts can support you in breaking down difficult subjects, building resilience, and developing important skills that will benefit you throughout your life.

It is important to remember that while AI can offer guidance and support, your curiosity and determination remain your greatest strengths. Rely on these qualities as you explore new topics and face challenges, knowing that your effort will shape your success.

Dear School Leaders,

Strong communication, effective planning, and staff support are essential for building and maintaining successful schools. These foundational elements create an environment where teachers, students, and administrators can thrive. The prompts provided in this playbook are designed to assist school leaders in a variety of important tasks, including developing policies, organizing events, and fostering meaningful engagement with the wider school community.

By thoughtfully, incorporating AI into everyday workflows, school leaders can streamline administrative responsibilities and enhance overall efficiency. This not only provides critical support to teachers, allowing them to focus on instruction and student relationships, but also helps strengthen connections with families and stakeholders, ensuring a collaborative and supportive educational experience for all.

How to Use This Playbook: Big-Picture Orientation

Purpose:

The main purpose of this section is to help readers understand the structure, goals, and pathways of the AI Prompt Playbook™. By reading this guide, educators and learners will gain insight into how to navigate and benefit from the book's resources.

Welcome to the AI Prompt Playbook™

This playbook serves as both a reference guide and a practical toolkit for educators and learners. Inside, you will discover a carefully curated collection of prompts that are designed to support teaching, learning, and effective school communication.

Structure of Each Prompt

- **The Why**: Each prompt begins with an explanation of the underlying principle or rationale, clarifying why the prompt is important.

- **The Goal**: This section defines what the user or teacher should aim to achieve by using the prompt.

- **The Toolkit/Steps**: Clear instructions or strategies are provided to guide you through the application of each prompt.

- **The Complete Output (where provided)**: Some prompts include ready-to-use examples to demonstrate their practical use.

- **Pro-Tips**: Additional advice is shared to ensure smooth use in the classroom or other real-world settings.

Ways to Use This Playbook:

There are three main ways to engage with this playbook:

- **Quick Reference**: Open the book to any prompt you need at the moment for immediate support.

- **Deep Study**: Explore entire categories systematically to foster professional growth and deepen your understanding.

- **Customization**: Adapt and remix prompts to fit your unique context, making them relevant for your specific needs.

How to Use These Prompts:
A Practical, Hands-On Guide

Purpose

This section provides teachers with a clear, step-by-step approach for applying individual prompts in their practice.

Step-by-Step Method for Using Prompts

- **Browse by Category**: Begin by selecting a category that matches your teaching objective, such as lesson design, assessment, engagement, or reflection. This helps ensure your chosen prompt is aligned with your goals.

- **Pick a Prompt**: Each prompt comes with essential metadata, including tone, format, role, and variables. Review these details to understand how the prompt can be effectively applied in your classroom or school environment.

- **Customize Variables**: Update the placeholders within the prompt to reflect your specific subject, grade level, or classroom context. Customizing in this way ensures that the prompt is relevant and meaningful to your students' needs.

- **Experiment and Adapt**: Use prompts as a foundation or starting point. Modify and refine the AI-generated outputs so they fit your teaching style, classroom culture, and students' learning objectives.

- **Ensure Ethical Use**: Keep in mind that AI tools should support—not replace—your professional judgment and empathy. Use prompts as helpful assistants, but always remain the primary decision-maker in your classroom.

Pro Tip:

Start small. Begin with just one or two prompts to see what resonates with your teaching style and classroom needs. After identifying what works best, gradually expand your use of prompts for greater impact.

Table of Contents

Category 4: Teacher Productivity & Support

Category 5: Student Growth & Development

Category 6: School Communication & Administration

Back Matter 443

Category 1: Lesson Design & Planning

Theme

This category focuses on the core building blocks of teaching, including lesson plans, unit outlines, worksheets, study guides, and classroom resources. It is designed to equip educators with structured tools that support the planning of effective and engaging lessons across various subjects and grade levels. By utilizing these resources, teachers can ensure their instruction is organized, purposeful, and tailored to promote student learning and achievement.

Category Goal

The primary aim of lesson design and planning is to help teachers save valuable preparation time while maintaining high standards for instructional quality. By focusing on alignment with learning objectives, lessons are structured to be interactive and impactful, ensuring that students remain engaged throughout the learning process. This deliberate approach provides students with a clear and guided pathway to achieving learning success.

Mini-Index: Lesson Design & Planning

Prompt	AI Role	Prompt Goal
Lesson Plan Designer	Curriculum Planner and Teacher Mentor	Provide complete, ready-to-use lesson plans
Unit Plan Designer	Curriculum Designer	Structure multi-week units with objectives and assessments
Worksheet & Exercise Creator	Resource Creator / Differentiation Specialist	Generate practice worksheets for specific topics
Study Guide Designer	Study Coach / Learning Strategist	Support student revision with structured study guides
Project Idea Generator	Project-Based Learning (PBL) Coach / Education Innovator	Offer creative project-based learning opportunities
Concept Explainer	Teacher / Learning Facilitator	Simplify complex concepts into age-appropriate explanations
Homework Assignment Generator	Teacher / Homework Designer	Create structured homework assignments with clear guidelines
Educational Game Generator	Gamification Coach / Teacher Support Specialist	Make learning interactive and memorable through games
Field Trip Idea Generator	Experiential Learning Architect / Field Trip Planner	Suggest real-world learning experiences tied to curriculum
Technology Integration Ideas	Educational Technology Integration Coach	Recommend ways to integrate digital tools in lessons

Detailed Prompts

Prompt: Lesson Plan Designer

Simple Prompt: "Act as an expert 'Curriculum Planner' and 'Teacher Mentor'. Create a complete lesson plan for [subject/topic] tailored to [grade/level], with purpose, standards, objectives, assessments, materials and resources, step-by-step procedures, and lesson plans."

- **Tone**: Structured, Professional
- **Format**: Lesson Plan (Objectives + Activities + Assessments)
- **Platform**: Classroom, Teacher Docs, LMS
- **AI Role**: Curriculum Architecture
- **Prompt Goal**: Provide a complete, ready-to-use lesson plan tailored to grade and subject
- **Tags**: #lessonPlan, #education, #planning
- **Prompt Variables**: {subject/topic}, {grade/level}, {objectives}, {activities}, {assessments}

Best Practice Tip: After generating a lesson plan, review and adapt activities to fit your classroom culture and student needs.

Enhanced Prompt:

Act as an expert 'Curriculum Planner' & 'Teacher Mentor'. Your task is to create a complete, ready-to-implement lesson plan for teaching [CONCEPT] to [GRADE LEVEL/COURSE]. The goal is to produce a structured plan that is engaging, aligns with backward design principles, incorporates varied learning modalities, and includes clear assessment and differentiation strategies to ensure all students can access and master the material.

The guide must be structured as follows:

1 **The Why it Matters**: A brief, impactful explanation of the lesson's purpose (e.g., "This lesson is designed using the 'Understanding by Design' framework, ensuring all activities and assessments are directly tied to the core learning objectives. This creates a coherent and purposeful learning experience that moves students beyond superficial knowledge to genuine understanding.").

2 **Standards & Alignment**: Identify 1-2 relevant curriculum standards (e.g., Common Core, NGSS, or state-specific standards) that this lesson addresses.

3 **Learning Objectives (The Destination)**: List 2-3 specific, measurable, and student-centered learning objectives framed as "Students will be able to..." (SWBAT) statements. (e.g., "SWBAT analyze the cause-and-effect relationship between historical events," "SWBAT solve two-step equations for an unknown variable," "SWBAT identify the main theme of a story and support it with textual evidence.").

4 **Assessment Evidence (How We Know They Know)**:

⋄ **Formative Assessment**: Describe 2-3 quick, ongoing checks for understanding to be used during the lesson (e.g., "Exit ticket," "Think-Pair-Share," "Thumbs up/down," "A quick poll on Kahoot!").

⋄ **Summative Assessment**: Describe the final product or task that will demonstrate mastery of the objectives at the end of the lesson or unit (e.g., "A short written response," "A completed

problem set," "A concept map," "A mini presentation.").

◇ **Success Criteria**: Provide a simple, single-point rubric or a list of "I can" statements
that clearly outline what a successful outcome
looks like for students.

5 **Materials & Resources**: A concise list of everything
needed for both teacher and students (e.g., "Projector,
whiteboard, student notebooks, linked video, graphic organizer handout, vocabulary list, math manipulatives").

6 **The Lesson Procedure**: A Step-by-Step Guide (Please
structure this with clear time allocations and dedicated
sections):

◇ **Phase 1**: Hook & Introduction (5-10 mins): An
engaging activity to activate prior knowledge,
spark curiosity, and introduce the lesson's goal.

◇ **Phase 2**: Direct Instruction / "I Do" (10-15
mins): Concise teacher-led modeling of the new
skill or concept. Think aloud and demonstrate
the process.

◇ **Phase 3**: Guided Practice / "We Do" (10-15
mins): A collaborative activity where the teacher
and students work through an example together,
with gradual release of responsibility.

◇ **Phase 4**: Independent Practice / "You Do" (10-
15 mins): Time for students to practice the skill
or apply the concept on their own or in small
groups.

◇ **Phase 5**: Closure & Reflection (5 mins): A structured wrap-up where students summarize their
learning, reflect on the objective, and/or com-

plete the formative assessment.

7 **Differentiation & Support (For All Learners)**:

◊ **For Support**: Provide 2-3 specific strategies (e.g., "Provide a pre-filled graphic organizer," "Use sentence starters," "Implement strategic small group instruction," "Offer a modified text version.").

◊ **For Extension**: Provide 2-3 specific strategies for students who need a challenge (e.g., "Offer an analysis of a more complex text," "Pose a 'what if' scenario," "Challenge them to apply the concept to a real-world problem.")

8 **Bonus**: Pro-Tip for Implementation:

◊ **Management Tip**: One piece of advice for smooth execution (e.g., "Have materials pre-bundled for each group to minimize transition time.").

◊ **Pacing Note**: A warning about a potential bottleneck (e.g., "The guided practice section is the most critical; do not rush it. Be prepared to flex your time here.").

The final output should be a complete plug-and-play lesson plan that any teacher can quickly personalize and execute with confidence, ensuring a high-impact learning experience for all students.

Complete Output:

Enhanced Lesson Plan Template (Backward Design Model)

Grade Level: [...] | Subject: [...] | Topic: [...] | Time Allotment [e.g., 60 Min

1. The "Why" It Matters: Overarching Goal: A brief explanation of the lesson's purpose and real-world relevance.

- **Example**: "This lesson moves beyond simple recall to help students understand the fundamental principle that [Big Idea]. This is crucial because it underpins [Real-World Application], empowering students to [Long-Term Benefit]."

2. Standards & Alignment

- **Content Standard(s)**: [List 1-2 primary content standards here, e.g., Common Core, NGSS, State Standard]
- **Skill Standard(s)**: [List 1-2 key skill standards here, e.g., Collaboration, Critical Thinking, Communication]

3. Learning Objectives (SMART Goals): Students will be able to (SWBAT):

- [Measurable Action Verb] + [Content] + [Context] (e.g., Analyze the author's use of symbolism in Chapter 3 to support a thematic claim.)
- [Measurable Action Verb] + [Content] + [Context] (e.g., Design a simple circuit that can power a light bulb using a battery, wires, and a switch.)
- [Measurable Action Verb] + [Content] + [Context] (e.g., Collaborate in a small group to solve a multi-step

word problem and explain the strategy used.)

4. Assessment Evidence: How Will You Know They Know?

- **Formative Assessment (During the Lesson)**:

    ◇ **Method**: [e.g., Exit ticket, Think-Pair-Share, Quick quiz on Kahoot! Observation of group work]

    ◇ **Purpose**: To check for understanding in real-time and inform immediate instructional adjustments.

- **Summative Assessment (End of Lesson/Unit)**:

    ◇ **Method**: [e.g., A short written response, a project, a presentation, a traditional quiz]

    ◇ **Success Criteria (For Students)**: Provide students with a simple rubric or checklist so they know what proficiency looks like.

        · **Example Checklist**:

            ▫ My answer includes a clear claim.

            ▫ I have provided at least two pieces of evidence.

            ▫ I have explained how my evidence supports my claim.

5. Materials & Resources

- **For Teacher**: [e.g., Projector, whiteboard, anchor chart, lesson slides, answer key]

- **For Students**: [e.g., Notebook, pen, handout, lab materials, device for online activity]

- **Digital Tools**: [e.g., Link to a specific simulation, YouTube video, Quizizz code]

6. Lesson Procedure: Step-by-Step Activities

Phase & Purpose	Teacher Activity & Student Activity	Time
Phase 1: Hook/ Launch (Activate prior knowledge, spark curiosity)	**Teacher**: Presents a compelling question, short video, image, or problem. **Students**: Think-Pair-Share their initial ideas or reactions.	~5-10 min
Phase 2: Direct Instruction/"I Do"(Model the new skill or concept)	**Teacher**: Explicitly models the thinking process. Thinks aloud. Introducing key vocabulary. **Students**: Actively listen, ask clarifying questions, take notes.	~10-15 min
Phase 3: Guided Practice / "We Do" (Collaborative practice with support)	**Teacher**: Facilitates; works through an example with the class. Provides scaffolds. **Students**: Work together to solve a problem or analyze a text with teacher guidance.	~10-15 min
Phase 4: Independent Practice/ "You Do" (Apply learn-ing individually)	**Teacher**: Circulates, provides targeted support, checks for understanding. **Students**: Work independently or in small groups to apply the new skill.	~15-20 min
Phase 5: Closure & Reflection (Solidify learning, assess understanding)	**Teacher**: Facilitates a share-out or administers an exit ticket. **Students**: Summarize their learning, reflect on the objective, complete exit ticket.	~5 min

7. Differentiation: Reaching Every Learner

- **For Support**: [e.g., Provide a word bank; use sentence starters; offer a graphic organizer; pre-teach key concepts in a small group; assign a peer partner.]

- **For Extension**: [e.g., Pose a "what if" scenario; provide a more complex text or problem; ask students to connect the concept to a different subject or current event; have them teach the concept to a peer.]

8. The "Why" Behind This Strategy

This backward design approach—starting with the end goal in mind—ensures that every activity and assessment is directly aligned to the core learning objectives. The "I Do, We Do, You Do" structure provides the necessary scaffolding to build student confidence and mastery, while built-in differentiation ensures all learners can access and engage with the content. This creates purposeful, efficient, and effective learning experience.

9. Bonus: Pro-Tip for Implementation

- **Print This Plan**: Have a copy with you to track timing.

- **Prep Materials in Advance**: Have handouts, tech links, and supplies ready to go to maximize instructional time.

- **Be Flexible**: If students struggle during "We Do," be prepared to slow down and model again. If they master it quickly, you can move on or provide deeper extension questions.

- **The Exit Ticket is Data**: Use the results from the exit ticket to group students for the next day's warm-up activity (e.g., who needs re-teaching, who is ready for enrichment).

To use this template, simply replace the bracketed [ ] information and examples with your specific topic, grade level, and standard. This structure provides a clear, rigorous, and flexible framework for any lesson.

Prompt: Unit Plan Designer

Simple Prompt: "Act as a 'Curriculum Designer (UbD framework)'. Create a complete, multi-week unit plan for [SUBJECT/TOPIC] designed for [GRADE LEVEL/COURSE] over [timeframe], including learning objectives, key instructional activities, and assessments."

- **Tone**: Comprehensive, Academic
- **Format**: Unit Plan (Multi-Week)
- **Platform**: Curriculum Documents, LMS
- **AI Role**: Curriculum Designer
- **Prompt Goal**: Provide teachers with structured unit-level planning to guide instruction
- **Tags**: #unitPlan, #curriculum, #planning
- **Prompt Variables**: {subject/topic}, {timeframe}, {objectives}, {activities}, {assessments}

Best Practice Tip: Use the AI-generated unit plan as a foundation, then personalize it with your own classroom pacing.

Enhanced Prompt:

Act as a Curriculum Designer (UbD framework). Your task is to create a complete, multi-week unit plan for [SUBJECT/TOPIC] designed for [GRADE LEVEL/COURSE]. The goal is to produce a robust, standards-aligned plan that is engaging, scaffolds learning effectively, and provides multiple ways for students to demonstrate mastery of big ideas and transferable skills.

The unit plan must be structured as follows:

1. **The "Why" it Matters**: A brief, impactful explanation of the unit's purpose and its place within the larger curriculum (e.g., "This unit on [TOPIC] is designed to help students understand the essential principle that [BIG IDEA]. This knowledge is foundational for [NEXT UNIT] and serves as a critical lens for understanding [REAL-WORLD APPLICATION].").

2. **Standards & Alignment**: List the 3-5 primary content standards (e.g., Common Core, NGSS, State Standards) and 1-2 key skill standards (e.g., Collaboration, Communication) this unit addresses.

3. **Big Ideas & Essential Questions**:

   ◇ **Big Ideas**: 2-3 overarching, transferable concepts students should retain long after the test (e.g., "Ecosystems are interdependent and dynamic.").

   ◇ **Essential Questions**: 2-3 open-ended, thought-provoking questions that frame the unit and spark inquiry (e.g., "How does a small change in an ecosystem create a ripple effect?").

4. **Learning Objectives (The Destination)**: A list of 3-5 specific, measurable, and student-centered "Students will be able to..." (SWBAT) statements that flow from the standards and big ideas.

5. **Final Assessment (Evidence of Understanding)**: Describe the authentic, summative performance task that will demonstrate mastery of the objectives and big ideas.

   ◇ **Task Description**: What will students do? (e.g., "Assume the role of an ecologist tasked with presenting a recovery plan for a threatened local ecosystem to a city council.")

◇ **Format**: What is the product? (e.g., "A written proposal with a visual model and a 5-minute oral presentation.")

◇ **Rubric Focus**: What are the key criteria for success? (e.g., "Content Accuracy, Use of Evidence, Proposed Solutions, Presentation Clarity").

6 **The Learning Plan**: A Scaffolded Progression Provide a day-by-day or week-by-week breakdown of the unit's arc. For each major phase, include:

◇ **Phase Title**: (e.g., "Phase 1: Building Foundational Knowledge")

◇ **Suggested Time Allocation**: (e.g., "Week 1")

◇ **Key Activities & Resources**: A bulleted list of 3-4 core learning experiences, with a mix of direct instruction, collaborative work, and independent practice. Note key resources (videos, texts, simulations).

◇ **Formative Assessments**: How you will check for understanding during this phase (e.g., "Exit ticket on key vocabulary," "Quiz on the steps of the process," "Rough draft of proposal outline.").

7 **Differentiation & Support (For All Learners)**:

◇ **For Support**: Provide 2-3 specific strategies (e.g., "Pre-teach key vocabulary," "Provide structured graphic organizers," "Offer sentence starters," "Create strategic small groups.")

◇ **For Extension**: Provide 2-3 specific strategies for students who need a challenge (e.g., "Research a case study of a failed ecosystem intervention and analyze the reasons," "Propose a innovative technological solution to the problem.")

8 **Materials & Resources**: A comprehensive, bulleted list of all needed items, categorized for clarity (e.g., Teacher Resources, Student Handouts, Technology Links, Manipulatives).

9 **The "Why" Behind This Strategy**: A concluding summary explaining how this backward-designed approach—starting with the end goal in mind—ensures coherence, alignment, and meaningful assessment, ultimately leading to deeper student learning and a more purposeful teaching experience.

10 **Bonus**: Pro-Tip for Teachers - Implementation:

⋄ **The Hook**: Start the unit with a provocative video, case study, or problem to ignite curiosity.

⋄ **Checkpoints**: Build in formal checkpoints for the summative task (e.g., proposal outline due, draft of presentation slides) to avoid last-minute work.

⋄ **Reflection**: End the unit with a metacognitive activity where students reflect on their learning journey in relation to the essential questions.

The final output should be a complete, plug-and-play unit plan that provides a clear roadmap for instruction, assessment, and differentiation, saving the teacher significant planning time.

Final Output

Grade Level:5 | Subject: Science | Units: Ecosystems

1. The "Why" it Matters: This unit moves beyond memorizing parts of a food web to understanding the profound interconnectedness and fragility of living systems. Students will learn that human actions have consequences and that they can be proactive problem-solvers in their own environment, a crucial mindset for informed citizenship.

2. Standards & Alignment:

- **NGSS 5-LS2-1**: Develop a model to describe the movement of matter among plants, animals, decomposers, and the environment.
- **NGSS 5-ESS3-1**: Obtain and combine information about ways individual communities use science ideas to protect the Earth's resources and environment.
- **Skill**: Collaborative Communication: Students will work in teams to negotiate ideas and present findings.

3. Big Ideas & Essential Questions:

- **Big Ideas**: Ecosystems are interdependent networks; human activity can disrupt or restore balance.
- **Essential Questions**: How do the choices we make today impact the world we live in tomorrow? What does it mean for an ecosystem to be "in balance"?

4. Learning Objectives (SWBAT):

- Identify and describe the roles of producers, consumers, and decomposers in a food web.
- Explain how energy and matter flow through an ecosystem.
- Analyze how a change in one population affects others

in the ecosystem.

- Propose a evidence-based solution to a real-world environmental threat.

5. Final Assessment: "Ecosystem Recovery Plan"

- **Task Description**: A local park's pond ecosystem is threatened by pollution and invasive species. Your team of ecologists must create a recovery plan.

- **Format**: A team-written proposal including a labeled diagram of the current and proposed food web, and a 5-minute presentation to the "city council" (the class).

- **Rubric Focus**: Accuracy of Models, Evidence-Based Reasoning, Feasibility of Solutions, Team Collaboration.

6. The Learning Plan: (Need to convert to a table with column names: Phase | Timeframe | Key Activities | Formative Assessments)

- **Phase 1**: What is an Ecosystem? (3-4 Days)

  ◇ **Activities**: Video hook (e.g., Our Planet), KWL chart, vocabulary sort, outdoor observation journaling.

  ◇ **Formative Assessment**: Exit ticket: "Draw and label the living and non-living parts of a mini-ecosystem."

- **Phase 2**: Energy Flow & Interdependence (4-5 Days)

  ◇ **Activities**: Create a physical food web with string and cards in groups, online simulation (e.g., Ecolab), decomposition jar experiment.

  ◇ **Formative Assessment**: Quiz on vocabulary and energy flow concepts.

- **Phase 3**: Human Impact & Case Studies (3-4 Days)

◇ **Activities**: Analyze case studies (e.g., reintro-duction of wolves to Yellowstone, local water quality issues), guest speaker from parks dept.

◇ **Formative Assessment**: "Think-Pair-Share: What was the most surprising impact from our case study?"

- **Phase 4**: Assessment & Application (4-5 Days)

  ◇ **Activities**: Introduce summative task, work time in teams, peer feedback on proposals, final presentations.

  ◇ **Formative Assessment**: Checkpoint 1: Team research outline. Checkpoint 2: Draft food web model.

7. Differentiation & Support:

- **For Support**: Vocabulary cards with images; pre-filled graphic organizers for food webs; option to present to teacher in a small group.
- **For Extension**: Research a global ecosystem resto-ration project; calculate the potential impact of their solution; create a public service announcement video instead of a presentation.

8. Materials & Resources:

- **Teacher**: Video clips, case study readings, quiz, rubric.
- **Student**: Science journals, vocabulary cards, string & index cards for food web activity, access to computers for research.
- **Technology**: Online ecosystem simulation, presenta-tion software (Google Slides/Canva).

9. The "Why" Behind This Strategy: This UbD plan ensures every activity directly builds the skills and knowledge needed for the authentic final assessment. It makes learning purposeful and visible to students, moving from basic knowledge to higher-order application and creation.

10. Bonus: Pro-Tip for Teachers:

- **Invite an Audience**: Have another class or school admin be the "city council" for the presentations to add authenticity.

- **Use a Single-Point Rubric**: This makes grading the project clearer and focuses feedback on growth.

Prompt: Worksheet & Exercise Creator

Simple Prompt: "Act as a 'Resource Creator' and 'Differentiation Specialist'. Generate 10 practice worksheet questions for [topic] with varying difficulty levels. Your task is to create a comprehensive, standards-aligned classroom worksheet on the topic of [TOPIC] for [GRADE LEVEL/COURSE]."

- **Tone**: Practical, Academic
- **Format**: Worksheet (10 Questions)
- **Platform**: Classroom, Homework, LMS
- **AI Role**: Resource Creator
- **Prompt Goal**: Provide ready-to-use worksheets for student practice
- **Tags**: #worksheets, #exercises, #practice
- **Prompt Variables**: {topic}, {difficultyLevels}, {grade/level}

Best Practice Tip: Use a mix of question types (MCQs, short answers, problems) to reinforce different skills.

Enhanced Prompt:

Act as a 'Resource Creator' and 'Differentiation Specialist'. Your task is to create a comprehensive, standards-aligned practice worksheet on the topic of [TOPIC] for [GRADE LEVEL/COURSE]. The goal is to generate a resource that effectively reinforces key concepts, provides tiered practice for diverse learners, and offers clear insights into student understanding, all while being efficient to grade.

The guide must be structured as follows:

1 **The "Why" it Matters**: A brief, impactful explanation of the core principle behind effective practice (e.g., "A strategically designed worksheet provides the necessary scaffolded repetition for skill automatization, while also serving as a formative assessment to identify students who need additional support or challenge.").

2 **Alignment to Learning Objectives**: State the primary learning objective this worksheet is designed to reinforce and assess (e.g., "Students will be able to solve two-step equations for an unknown variable.").

3 **The Worksheet**: Tiered Practice Provide a set of [NUMBER] exercises. The worksheet must be intentionally sequenced to build complexity

   ◊ **Part 1**: Foundational Practice (Approx. 30%): Simple exercises that build confidence and assess basic comprehension of the core skill or concept.

   ◊ **Part 2**: Application Practice (Approx. 50%): More complex exercises that require students to apply the skill in standard contexts.

   ◊ **Part 3**: Challenge/Extension (Approx. 20%): One or two problems that push students to synthesize this skill with prior knowledge or apply it in a novel, real-world context.

4 **Breakdown for Each Exercise**: For every exercise, include:

   ◊ **Exercise Number & Text**: The problem or question presented clearly.

   ◊ **Exercise Type**: Label the format (e.g., Fill-in-the-Blank, Short Answer, Problem-Solving,

Matching, Diagram Labeling).

◇ **Target Skill**: Specify the micro-skill or piece of knowledge this exercise practices.

◇ **Answer Key**: The correct answer or solution.

5 **The "Why" Behind This Strategy**: A concluding summary explaining how this tiered approach supports mastery learning, allows for natural differentiation (e.g., "All students attempt Part 1 & 2, while Part 3 is optional enrichment"), and provides the teacher with a clear picture of where each student is in their learning journey.

6 **Bonus**: Pro-Tip for Implementation & Feedback: Add a short section with practical advice for the teacher:

◇ **Implementation**: How to use the worksheet effectively (e.g., "Use Part 1 as a warm-up review, assign the rest for independent practice," "Have students self-check Part 1 using a provided answer key to promote self-assessment.").

◇ **Error Analysis Guide**: A note on common errors to look for while grading & what they indicate (e.g., "Mistakes in Part 2 often indicate a misunderstanding of [specific rule], which will require a small group re-teach.")

The final output should be a complete, plug-and-play practice packet that is easy to administer, grade, and use to inform instructional next steps.

Final Output

Worksheet: Mastering Fractions

Grade Level: 6 | Subject: Math | Units: Fractions

Objective: Students will be able to simplify fractions, perform operations (add, subtract, multiply, divide), and solve real-world problems involving fractions.

1. The "Why" It Matters

Fractions are everywhere — from dividing pizza to comparing test scores. This worksheet builds confidence with foundational fraction skills, applies them to real-world scenarios, and challenges students to think critically.

2. Alignment to Learning Objective

SWBAT: Solve fraction problems with accuracy, apply fraction operations in real-world contexts, and explain their reasoning.

Part 1: Foundational Practice (Confidence Builders)

(Basic comprehension & recall – ~30%)

1 **Simplify**: $1218\frac{12}{18}1812 \to$ _____ (Type: Simplify fraction | Target Skill: Simplifying by GCF)

2 **Convert**: Write $34\frac{3}{4}43$ as a decimal. (Type: Short Answer | Target Skill: Fraction-decimal conversion)

3 **Compare**: Which is greater: $58\frac{5}{8}85$ or $34\frac{3}{4}43$? (Type: Comparison | Target Skill: Benchmark fractions)

Part 2: Application Practice (Core Skill)

(Standard exercises in context – ~50%)

4 **Add**: $25+310=\frac{2}{5} + frac{3}{10} =52+103=$ ___ (Type: Problem-Solving|Target Skill: Finding common denominators)

5 **Subtract**: $78-512=\frac{7}{8} - \frac{5}{12} =87-125 =$ _____ (Type: Problem-Solving | Target Skill: Subtract

unlike fractions)

6 **Multiply**: 34×25=\frac {3 {4 times frac {2}{5} = 43 ×52= _____ (Type: Problem-Solving | Target Skill: Fraction multiplication)

7 **Divide**: 56÷23=\frac{5}{6} div \frac{2}{3} = 65÷32 = ____ (Type: Problem-Solving | Target Skill: Fraction division / reciprocal)

8 **Word Problem**: Sarah ate 25\ frac {2 {5}52 of a cake and John ate 14\frac{1}{4}41 How much of the cake did they eat together? (Type: Word Problem | Target Skill: Adding fractions in context)

Part 3: Challenge & Extension

(Novel / Real-world applications – ~20%)

9 **Mixed Operation Challenge**: 38+23×910 =\ frac {3}{8} + \frac{2}{3} \times \frac {9}{10} =83+32×109 = _____ (Type: Multi-step | Target Skill: Order of operations with fractions)

10 **Real-World Challenge**: A recipe calls for 34\frac{3}{4}43 cup of sugar. If you want to make 23\frac{2}{3}32 of the recipe, how much sugar do you need? (**Type**: Real-World Application | Target Skill: Scaling with fractions)

Answer Key

1   23\frac{2}{3}32

2. 0.75

3. 34\frac{3}{4}43

4. 710\frac{7}{10}107

5. 1124\frac{11}{24}2411

6. $3\frac{3}{10}$

7. $5\frac{5}{4}$ or $11\frac{1}{4}$

8. $13\frac{13}{20}$

9. $83\frac{83}{120}$

10. $1\frac{1}{2}$ cup

Teacher Notes

- **Implementation**:

  ◇ Use Part 1 as a warm-up or quick review.

  ◇ Assign Part 2 for guided or independent practice.

  ◇ Treat Part 3 as enrichment, group challenges, or homework extension.

- **Error Analysis Tips**:

  ◇ Watch for denominator mistakes in addition/subtraction (students forgetting to find LCM).

  ◇ In division, many forget to "flip the second fraction."

  ◇ Word problems often reveal misunderstandings of "parts of a whole."

- **Differentiation**:

  ◇ Provide fraction strips or visuals for struggling learners.

  ◇ Encourage advanced students to create their own fraction word problems.

This format is plug-and-play: swap {topic}, {grade}, or {number of exercises} to instantly create new worksheets.

Prompt: Study Guide Designer

Simple Prompt: "Act as a 'Study Coach' and 'Learning Strategist'. Your task is to create a student-friendly, strategic study guide for [SUBJECT], specifically covering [CHAPTERS/UNITS] for a [GRADE LEVEL/COURSE] student".

- **Tone**: Concise, Supportive
- **Format**: Study Guide Document
- **Platform**: Student Handouts, Revision Notes
- **AI Role**: Study Coach/Learning Strategist
- **Prompt Goal**: Help students revise and retain essential knowledge
- **Tags**: #studyGuide, #revision, #learning
- **Prompt Variables**: {topic}, {grade/level}, {studyGoals}

Best Practice Tip: Encourage students to add their own notes to personalize the guide.

Enhanced Prompt:

Act as a 'Study Coach' and 'Learning Strategist'. Your task is to create a comprehensive, strategic study guide for [SUBJECT], specifically covering [CHAPTERS/UNITS] for a [GRADE LEVEL/COURSE] student. The goal is to produce a resource that moves beyond simple content summarization to teach students how to study effectively by focusing on active recall, spaced practice, and self-assessment.

The guide must be structured as follows:

1. **The "Why" it Matters**: A brief, impactful explanation of the core principle behind effective studying (e.g., "Passively re-reading notes is one of the least effective study methods. This guide is designed around the science of learning, prioritizing active retrieval and the identification of knowledge gaps to maximize retention and understanding.").

2. **Overarching Learning Objectives**: List the 3-5 most critical, big-picture learning goals a student should master after studying these chapters (e.g., "Explain the causes and effects of World War I," "Apply the laws of thermodynamics to real-world systems," "Analyze the major themes in Shakespeare's Macbeth.").

3. **The Strategic Study Guide**: The guide should be organized by chapter/unit and must include the following dedicated sections for each:

   ◇ **Chapter Summary**: A concise, student-friendly overview of the chapter's core narrative or purpose (2-3 sentences).

   ◇ **Key Concepts & Terminology**: A bulleted list of 3-5 essential ideas and their definitions. This is the "what you need to know" list.

   ◇ **Active Recall Tasks**: This is the most critical section. Provide specific, action-oriented tasks for the student to complete without looking at their notes (e.g., "Explain the process of photosynthesis to a friend without using your textbook," "Draw and label a diagram of a cell from memory," "List the five key steps of the scientific method.").

   ◇ **Practice Questions**: A mix of 3-5 question types (e.g., multiple-choice, short answer, diagram interpretation) that test application and analysis, not just recall. Important: Do not provide the answers here.

4 **The Answer Key & Self-Assessment Guide**: A separate, clearly labeled section containing:

⋄ **Answers**: The correct answers and solutions to all practice questions.

⋄ **Self-Assessment Tip**: For tougher questions, include a note on what to do if the student got it wrong (e.g., "If you missed this, re-read page 42 on [concept] and try a similar problem.").

5 **The "Why" Behind This Strategy**: A concluding summary explaining how this active, self-testing approach builds stronger neural pathways, reduces study time by focusing on weaknesses, and builds student confidence through proven methods.

6 **Bonus**: Pro-Tip for the Student - How to Use This Guide: Add a short section with actionable study advice.

⋄ **The Study Schedule**: A suggested plan (e.g., "Study one chapter per day using this guide. The day before the test, only review the Key Concepts and the questions you got wrong.").

⋄ **Effective Study Techniques**: Recommend specific methods (e.g., "Use the Feynman Technique with the Key Concepts: try to explain them in simple terms," "Create flashcards for the Terminology list," "Teach the Chapter Summary to someone else.").

The final output should be a complete, plug-and-play strategic study plan that empowers a student to study smarter, not harder, and to take ownership of their learning.

Complete Output

Strategic Study Guide: Cell Biology

Grade Level: 9 | Subject: Science | Units: Cell Structure & Function, Cell Processes

1. The "Why" It Matters

Understanding cell biology is crucial because cells are the building blocks of life. Mastering how cells function provides the foundation for future topics like genetics, human anatomy, and disease. This guide uses active recall, self-testing, and spaced practice to help you study smarter, not harder.

2. Overarching Learning Objectives

By the end of this study, students will be able to:
- Identify and describe the structure and function of major cell organelles.
- Differentiate between prokaryotic and eukaryotic cells.
- Explain processes such as osmosis, diffusion, and active transport.
- Analyze how cell functions contribute to the survival of the organism.

3. The Strategic Study Guide:

Chapter/Unit 1: Cell Structure: Chapter Summary: Cells are the smallest units of life. Organelles inside cells each have specific functions that keep the cell alive and working.

- **Key Concepts & Terminology**:

  ◇ Nucleus – control center, contains DNA.

  ◇ Mitochondria – powerhouse, produces ATP energy.

  ◇ Ribosomes – protein synthesis.

  ◇ Cell Membrane – selectively permeable barrier.

  ◇ Prokaryote vs. Eukaryote – with/without a nucleus.

Active Recall Tasks:

- Draw and label a plant and animal cell from memory.
- **Without notes**: explain why mitochondria are called the powerhouse of the cell.
- List three differences between prokaryotic and eukaryotic cells.

Practice Questions:

- **Multiple Choice**: Which organelle is responsible for protein synthesis?
- **Short Answer**: Compare plant and animal cells in 2 differences and 2 similarities.
- **Diagram Labeling**: Label 5 organelles on a blank cell diagram.
- **Explain**: Why is the cell membrane considered selectively permeable?

Chapter/Unit 2: Cell Processes

Chapter Summary: Cells maintain balance by moving substances in and out. They obtain energy through respiration and photosynthesis (plants).

Key Concepts & Terminology:

- Diffusion – movement from high → low concentration.
- Osmosis – diffusion of water across a membrane.
- Active Transport – energy-requiring movement against the gradient.
- Cell Respiration – glucose → energy (ATP).
- Photosynthesis – plants convert sunlight → glucose.

Active Recall Tasks:

- **Teach a peer**: explain osmosis in your own words.
- **From memory**: write the equation for photosynthesis and respiration.
- Explain the difference between passive and active transport without looking at notes.

Practice Questions:

1 **Multiple Choice**: Which process requires energy?

2 **Short Answer**: Explain why osmosis is essential for maintaining homeostasis.

3 **Problem**: A cell is placed in salty water. Predict what happens and explain why.

4 **Diagram Task**: Draw and annotate the steps of photosynthesis.

4. Answer Key & Self-Assessment Guide

Sample Answers:

- **Cell Structure Q1**: Ribosomes.
- **Q2**: Similarities – both have cell membranes, cytoplasm. Differences – plants have chloroplasts & cell walls, an-

imals do not.

- **Q4**: The membrane only allows certain molecules through (like a filter), keeping balance.

- **Cell Processes Q1**: Active transport.

- **Q2**: Osmosis balances water in/out of the cell to avoid dehydration or bursting.

- **Q3**: The cell shrinks (water moves out) – hypertonic solution.

- **Q4**: Photosynthesis: $6CO_2 + 6H_2O \rightarrow C_6H_{12}O_6 + 6O_2$ (requires sunlight, chlorophyll).

Self-Assessment Tip: If you missed a question, re-check diagrams, vocabulary, or equations in your notes. Redo the task from memory without looking at the answer until you can recall it independently.

5. The "Why" Behind This Strategy

This guide emphasizes active recall (retrieving from memory), practice questions (application), and spaced repetition (reviewing concepts over time). This approach builds stronger memory pathways, helps you focus on weak spots, and boosts confidence for tests.

6. Bonus: Pro-Tip – How to Use This Guide

- **Study Schedule**:

  ◇ **Day 1**: Chapter 1 (structure).

  ◇ **Day 2**: Chapter 2 (processes).

  ◇ **Day 3**: Practice both sets of recall tasks.

  ◇ **Day 4**: Re-do only the questions you missed.

  ◇ **Day 5**: Full review before test.

- **Techniques**:

  ◇ Use flashcards for organelles and processes.

  ◇ Practice drawing diagrams repeatedly until you can do it from memory.

  ◇ Teach the processes (osmosis, respiration) aloud as if you're the teacher.

Complete Output: A ready-to-use study guide for Grade 9 Science (Cell Biology), fully structured and student-friendly.

Prompt: Project Idea Generator

Simple Prompt: "Act as a 'Project-Based Learning (PBL) Coach' and 'Education Innovator'. Your task is to create a teacher-ready menu of authentic classroom project ideas for [SUBJECT], specifically designed for [GRADE LEVEL/COURSE]."

- **Tone**: Creative, Practical
- **Format**: Project Idea List
- **Platform**: Class Projects, Group Work
- **AI Role**: Education Innovator
- **Prompt Goal**: Provide project ideas that foster creativity and applied learning
- **Tags**: #projects, #learningByDoing, #creativity
- **Prompt Variables**: {topic}, {grade/level}, {project-Duration}

Best Practice Tip: Ensure project ideas are flexible enough for different class sizes and resources.

Enhanced Prompt:

Act as a 'Project-Based Learning (PBL) Coach' and 'Education Innovator'. Your task is to create a comprehensive menu of authentic project ideas for [SUBJECT], specifically designed for [GRADE LEVEL/COURSE]. The goal is to generate ideas that are not only engaging but are also rooted in real-world application, requiring students to synthesize knowledge, cultivate 21st-century skills, and create a meaningful public product.

The guide must be structured as follows:

1 **The "Why" it Matters**: A brief, impactful explanation of the core principle behind Project-Based Learning (e.g., "Authentic projects transform students from passive recipients of information into active creators of knowledge. They provide a purpose for learning, fostering deeper understanding, resilience, and the ability to apply skills in novel contexts—preparing students for real-world challenges.").

2 **Overarching Learning Objectives**: List the 2-3 primary skills and knowledge standards this set of projects is designed to assess (e.g., "Knowledge: Students will demonstrate understanding of the principles of ecology. Skills: Students will collaboratively design a solution, analyze data, and present findings to an audience.").

3 **The Project Menu**: A Spectrum of Authentic Ideas Provide a list of 5-7 distinct project ideas. Ensure there is a deliberate mix of the following types to cater to diverse strengths & interests:

   ◇ **Design & Build**: Creating a physical or digital model, prototype, or product.

   ◇ **Research & Report**: Conducting deep inquiry and presenting findings in a professional format.

   ◇ **Persuasive & Advocate**: Crafting a campaign, speech, or proposal to influence an audience.

   ◇ **Creative & Narrative**: Using storytelling, art, or media to convey understanding.

   ◇ **Collaborative & Simulation**: Role-playing a real-world scenario or working as a team to solve a complex problem.

4 **Breakdown for Each Project Idea**: For every idea,

include:

- ◇ **Project Title**: A catchy, student-friendly name.

- ◇ **Driving Question**: The open-ended, complex question that frames the project and guides student inquiry (e.g., "How can we, as urban planners, design a green space to reduce our city's carbon footprint and improve community health?").

- ◇ **Core Concept**: A 1-2 sentence description of the project's goal and final product.

- ◇ **The "Why" it's Important**: A brief explanation of the long-term skill development and real-world relevance (e.g., "This project develops systems thinking, ethical reasoning, and persuasive communication skills, mirroring the work of real environmental scientists and policymakers.").

- ◇ **Key Components & Steps**: A bulleted list of 3-4 essential tasks or milestones (e.g., 'Conduct research on local ecosystems,' 'Interview a community mem-ber,' 'Build a scaled model,' 'Present findings to a panel').

5 **The "Why" Behind This Strategy**: A concluding summary explaining how this menu of authentic, student-driven projects increases intrinsic motivation, assesses higher-order thinking skills, and creates a classroom culture of innovation and critical thinking that far surpasses the impact of traditional assessments.

6 **Bonus**: Pro-Tip for Implementation - The PBL Framework: Add a short section with actionable advice for the teacher to ensure project success.

- ◇ **Student Choice & Voice**: "Allow students to

choose their project or adapt the product, increasing ownership and engagement."

◇ **Scaffolding & Milestones**: "Break the project into phases with clear deadlines for proposals, drafts, and revisions to manage time and provide formative feedback."

◇ **Public Product**: "Whenever possible, have students present their work to an audience beyond the classroom (e.g., peers, parents, community experts) to add authenticity and stakes."

◇ **Reflection**: "Build in a reflection component for students to metacognitively assess their learning process and growth."

The final output should be a complete plug-and-play project menu that provides a clear path for implementation, from the initial driving question to the final public product, empowering students to do meaningful work.

Complete Output

Project Menu: Renewable Energy Innovations

Subject: Science | Grade Level: 8 | Duration: 4–6 weeks

1. The "Why" it Matters

Authentic projects transform students from passive recipients of information into active creators of knowledge. By tackling renewable energy challenges, students engage in real-world problem solving, develop critical thinking, and learn how science directly impacts communities and the planet.

2. Learning Objectives

- **Knowledge**: Students will demonstrate understanding of renewable energy sources and their applications.

- **Skills**: Students will collaborate, research, design solutions, and present their findings to authentic audiences.

3. Project Menu: Authentic Ideas

- **Design a Solar-Powered Model House**

  ◇ **Driving Question**: How can we design a home that runs entirely on renewable energy?

  ◇ **Core Concept**: Students build a scaled model of a house powered by solar panels.

  ◇ **Why It's Important**: Encourages systems thinking and engineering design.

  ◇ **Key Steps**:

    - Research solar power basics.

    - Sketch energy-efficient house plans.

    - Build a prototype with solar kits.

    - Present design and test results to peers.

- **Community Energy Awareness Campaign**

  ◇ **Driving Question**: How can we raise awareness about renewable energy in our community?

  ◇ **Core Concept**: Students create a multimedia campaign (videos, posters, social media).

  ◇ **Why It's Important**: Builds persuasive communication and advocacy skills.

  ◇ **Key Steps**:

- Survey local energy usage habits.

- Create informational content.

- Launch campaign (school bulletin boards or local library).

- Reflect on impact.

- **Energy Audit of the School**

  ◇ **Driving Question**: How energy-efficient is our school, and how can we improve?

  ◇ **Core Concept**: Students conduct a school energy audit and propose solutions.

  ◇ **Why It's Important**: Links classroom science to immediate, practical change.

  ◇ **Key Steps**:

    - Gather data on lighting, heating, and water use.

    - Identify energy waste.

    - Propose realistic improvements.

    - Present findings to school leadership.

- **Storytelling for Change: The Future of Energy**

  ◇ **Driving Question**: What would a 100% renewable-energy-powered world look like?

  ◇ **Core Concept**: Students write a narrative, create a comic, or produce a short film.

  ◇ **Why It's Important**: Encourages creativity and future-oriented thinking.

◇ **Key Steps**:

- Research renewable technologies.

- Create a fictional story showcasing their impact.

- Share stories in a class "Energy Futures Festival."

- **Debate: Renewable Energy vs. Fossil Fuels**

  ◇ **Driving Question**: Should our society transition fully to renewable energy within 30 years?

  ◇ **Core Concept**: Structured debate requiring evidence-based arguments.

  ◇ **Why It's Important**: Builds critical thinking, argumentation, and listening skills.

  ◇ **Key Steps**:

    - Divide into pro/con teams.

    - Research evidence and statistics.

    - Conduct formal debate.

    - Reflect in a written piece.

- **Prototype Challenge: Wind Turbine Design**

  ◇ **Driving Question**: What design makes a wind turbine most efficient at generating power?

  ◇ **Core Concept**: Students design and test small wind turbine models.

  ◇ **Why It's Important**: Hands-on engineering experience with data collection.

◇ **Key Steps**:

- Research turbine blade designs.

- Build prototypes from simple materials.

- Test with fans, measure output.

- Present efficiency findings.

4. Why This Strategy Works: This menu offers diverse entry points—engineering, advocacy, storytelling, research—so every student can shine. It promotes higher-order thinking, ensures real-world connections, and results in meaningful public products.

5. Bonus: Pro-Tip for Teachers – The PBL Framework

- **Student Choice & Voice**: Let students pick from the menu or adapt a project.

- **Scaffolding & Milestones**: Break projects into weekly checkpoints.

- **Public Product**: Have students present at a school-wide "Energy Expo."

- **Reflection**: End with a student reflection on skills gained and real-world impact.

Prompt: Concept Explainer

Simple Prompt: "Act as a 'Teacher' and 'Learning Facilitator'. Your task is to create a clear, multi-sensory explanation toolkit for the concept of [CONCEPT] tailored specifically to a [GRADE LEVEL] student".

- **Tone**: Simple, Clear
- **Format**: Explanation Text
- **Platform**: Classroom, Study Notes
- **AI Role**: Teacher / Learning Facilitator
- **Prompt Goal**: Break down complex ideas into digestible concepts
- **Tags**: #concepts, #simplification, #teaching
- **Prompt Variables**: {complexConcept}, {grade/level}, {examples}

Best Practice Tip: Pair the explanation with a diagram or analogy for better retention.

Enhanced Prompt:

Act as a 'Teacher' and 'Learning Facilitator'. Your task is to create a comprehensive, multi-sensory explanation kit for the concept of [CONCEPT] tailored specifically to a [GRADE LEVEL] student. The goal is to break down this complex idea into its core components using evidence-based strategies that make it accessible, memorable, and engaging for all learners.

The guide must be structured as follows:

1   **The "Why" it Matters**: A brief, impactful explanation of the core principle behind effective knowledge transfer (e.g., "Students grasp difficult concepts not through repetition, but when information is chunked, connected to prior knowledge, and presented through multiple modalities. This approach builds robust mental models instead of fragile, superficial understanding.").

2   **Core Learning Objective**: State the one, fundamental takeaway a student should have after this explanation (e.g., "The student will be able to define [CONCEPT] in their own words and provide a novel example.").

   ◇   **The Explanation**: Chunked and Scaffolded

   ◇   **Core Idea**: Start with a single, simple sentence that captures the absolute essence of the concept.

   ◇   **Breaking it Down**: Explain the concept in 2-3 clear, logical steps or components. Use bullet points for clarity. Avoid jargon; use language appropriate for the grade level.

   ◇   **Key Terms**: Bold any essential vocabulary words and provide a super-simple definition for each within the flow of the explanation.

3   **The "Aha!" Moment Tools**: Provide two powerful tools to create clarity:

   ◇   **The Analogy**: A relatable, high-quality analogy that connects the new concept to something the student already knows from their world (e.g., "Understanding a cell is like understanding a factory. The nucleus is the boss's office, the mitochondria are the power generators, and the endoplasmic reticulum is the assembly line.").

   ◇   **The Real-World Example**: A simple, concrete

example of how this concept appears in everyday life, technology, or nature.

4 **Check for Understanding & Application**: Include a multi-tiered approach to assess comprehension:

◇ **Quick Check Question**: A simple recall question to verify basic understanding (e.g., "So, in our analogy, what part of the cell acts like the boss?").

◇ **Application Challenge**: A question that requires the student to apply the concept in a new way (e.g., "If a cell was a city, what parts would the power plant, the library, and the delivery trucks be? Why?").

5 **The "Why" Behind This Strategy**: A concluding summary explaining how this multi-faceted approach (chunking, analogies, examples) respects cognitive load theory, builds bridges in the brain for better recall, and empowers the student to be an active participant in their learning.

6 **Bonus**: Pro-Tip for the Teacher - Make it Stick: Add a short section with an actionable strategy for reinforcing the concept.

◇ **Visual Aid Idea**: A suggestion for a quick drawing or diagram (e.g., "Draw the factory analogy on the whiteboard as you explain.").

◇ **Kinesthetic Hook**: A simple physical movement or hand gesture that represents the concept (e.g., "Use your arms to show the push and pull of opposing forces.").

◇ **Extension for Early Finishers**: A curious question or a quick "what if" scenario to push students who have already grasped the concept.

The final output should be a complete, plug-and-play explanation kit that empowers a teacher to deliver a concept with clarity and confidence, turning a potential point of frustration into a moment of breakthrough for the student.

Complete Output

Grade Level: 7 | Subject: Science | Concept: Photosynthesis

1. The "Why" it Matters

Students grasp difficult concepts not through repetition, but when information is chunked, connected to prior knowledge, and presented through multiple senses. Photosynthesis is a foundation for understanding how energy flows through ecosystems, and making it concrete helps students see its importance in their daily lives.

2. Core Learning Objective

SWBAT (Students Will Be Able To): Explain photosynthesis in their own words and provide an example of how plants use sunlight to create energy.

3. The Explanation (Chunked & Scaffolded)

- **Core Idea**: Photosynthesis is the process plants use to turn sunlight into food.
- **Breaking it Down**:

    1 Plants take in carbon dioxide (CO_2) from the air and water (H_2O) from the soil.

    2. With the help of sunlight and chlorophyll (the green pigment in leaves), plants change these

into glucose (sugar) and oxygen (O_2).

3. The sugar is used for energy and growth, while oxygen is released into the air.

- **Key Terms**:

  ◇ **Chlorophyll**: The green pigment in plants that captures sunlight.

  ◇ **Glucose**: A type of sugar that plants make for food.

  ◇ **Oxygen**: A gas plants release that humans and animals breathe.

4. The "Aha!" Moment Tools

- **Analogy**: Photosynthesis is like a solar-powered kitchen. The sunlight is the stove, the ingredients are water and carbon dioxide, and the meal made is sugar (food for the plant), with oxygen as the leftover steam released.

- **Real-World Example**: When you see a forest, you're looking at the planet's "oxygen factory." Every tree and leaf is working like a kitchen, making food for itself and releasing the oxygen we breathe.

5. Check for Understanding & Application

- **Quick Check Question**: What gas do plants release into the air after photosynthesis?

- **Application Challenge**: Imagine Earth without photosynthesis. What would happen to animals, including humans? Why?

6. The "Why" Behind This Strategy

By combining chunked steps, analogies, and real-life connections, this explanation reduces cognitive overload and helps students form strong mental models. The quick checks ensure

1 **The "Why" it Matters**: A brief, impactful explanation of the modern principle behind effective homework (e.g., "The purpose of homework is not to keep students busy; it is to strategically extend learning beyond the classroom. Effective homework provides deliberate practice, fosters independent learning habits, and offers the teacher crucial diagnostic data on student understanding to inform future instruction.").

2 **Alignment to Learning Objectives**: State the 2-3 primary skills or knowledge standards this set of assignments is designed to reinforce and assess (e.g., "To apply the formula for calculating area to real-world scenarios," "To identify the main idea and supporting details in a complex text," "To formulate a hypothesis based on observable phenomena.").

3 **The Assignment Menu**: Purposeful & Varied Tasks Provide a list of 5-7 distinct assignment ideas. Ensure there is a deliberate mix of the following types to serve different purposes and learning styles:

  ◊ **Practice & Reinforcement**: Short, focused tasks to solidify a core skill.

  ◊ **Preparation & Preview**: Tasks that build background knowledge for upcoming lessons.

  ◊ **Application & Creation**: Tasks that require applying knowledge to a new context or creating a product.

  ◊ **Reflection & Metacognition**: Tasks that prompt students to think about their own learning process.

4 **Breakdown for Each Assignment Idea**: For every idea, include:

⬦ **Assignment Title**: A clear, student-friendly name for the task.

⬦ **Task Description**: A 1-2 sentence description of what the student needs to do, phrased as direct instructions.

⬦ **Estimated Time**: A realistic time commitment (e.g., "15-20 minutes").

⬦ **The "Why" it's Important**: A brief explanation of the specific skill it targets and its long-term benefit (e.g., "This task moves beyond simple recall to force synthesis of concepts, a critical skill for essay writing and problem-solving.").

⬦ **Submission Format & Feedback Focus**: How the student should submit it and what the teacher will look for (e.g., "Submit as a shared Google Doc; focus feedback on the clarity of the thesis statement.").

5 **The "Why" Behind This Strategy**: A concluding summary explaining how this menu of purposeful, time-bound assignments increases student buy-in, provides a more accurate picture of individual and class-wide understanding, and respects student well-being by replacing volume with value.

6 **Bonus**: Pro-Tip for Teachers - The Feedback Loop: Add a short section with actionable advice for implementation.

⬦ **The 5-Minute Rule**: "If an assignment cannot be designed to be completed in a focused, age-appropriate time window (e.g., 20 mins for MS, 30-45 for HS), it should be reconsidered."

⬦ **Closing the Loop**: "Always incorporate a brief

review of homework at the start of the next class. This validates the effort, addresses common errors quickly, and makes the homework meaningful. Use a quick strategy like 'Think-Pair-Share' on a key problem."

◇ **Differentiation**: "Offer choice where possible. For example, allow students to choose 3 out of 5 practice problems to complete, or choose between writing a paragraph or drawing a diagram to explain a concept."

The final output should be a complete, plug-and-play homework menu that is easy to integrate into a unit plan, provides clear value to both student and teacher, and transforms homework from a chore into a meaningful learning extension.

Complete Output

Homework Assignment Menu: Linear Equations

Grade Level: 8 | Subject: Math | Unite: Linear Equations:

1. The "Why" it Matters

The purpose of homework is not to keep students busy; it is to strategically extend learning beyond the classroom. Effective homework offers deliberate practice, fosters independence, and gives teachers data to adjust instruction. For linear equations, meaningful assignments strengthen algebraic fluency while connecting math to real-life problem solving.

2. Alignment to Learning Objectives

This homework menu reinforces:

- **Skill 1**: Solving one and two-step linear equations.

- **Skill 2**: Applying linear equations to real-world problems.

- **Skill 3**: Reflecting on mathematical strategies and errors.

3. The Assignment Menu (Purposeful & Varied Tasks)

- **Practice Drill**: Equation Warm-Up

  ◇ **Task**: Solve 10 linear equations (mix of one-step and two-step).

  ◇ **Estimated Time**: 15 minutes.

  ◇ **Why It's Important**: Builds speed and accuracy with fundamental algebraic manipulations.

  ◇ **Submission Format & Feedback**: Paper submission; teacher checks for correct methods and clear steps.

- **Preview Task**: Equation Word Problems

  ◇ **Task**: Read 3 short word problems involving unknowns (no solving required). Write the linear equation that matches each problem.

  ◇ **Estimated Time**: 15 minutes.

  ◇ **Why It's Important**: Prepares students for upcoming lessons by focusing on equation setup.

  ◇ **Submission Format & Feedback**: Submit equations only; feedback highlights translation of words into symbols.

- **Real-Life Application**: My Budget Equation

  ◇ **Task**: Write and solve an equation for a simple real-life situation (e.g., allowance savings, cell phone bill, or movie tickets).

◊ **Estimated Time**: 20 minutes.

◊ **Why It's Important**: Connects classroom math to personal experiences, deepening relevance and retention.

◊ **Submission Format & Feedback**: Submit short written scenario, equation, and solution. Feedback focuses on creativity and accuracy.

- **Reflection Task**: "My Strategy"

  ◊ **Task**: Solve 2 equations. Then write 3–4 sentences explaining the steps you took and why.

  ◊ **Estimated Time**: 10–15 minutes.

  ◊ **Why It's Important**: Encourages metacognition, helping students articulate their problem-solving process.

  ◊ **Submission Format & Feedback**: Teacher responds with one strength and one suggestion for clarity.

- **Challenge Problem**: Puzzle Equation

  ◊ **Task**: "The sum of three consecutive odd numbers is 75. Find the numbers."

  ◊ **Estimated Time**: 15 minutes.

  ◊ **Why It's Important**: Develops problem-solving resilience and synthesis of algebra skills.

  ◊ **Submission Format & Feedback**: Solution with work shown; teacher provides enrichment feedback for advanced problem-solvers.

4. The "Why" Behind This Strategy: This menu approach

balances practice with creativity and reflection. It gives students choice, respects time, and offers the teacher varied evidence of understanding. By including reflection and real-life connections, homework becomes more meaningful than rote drills.

5. Bonus: Pro-Tip for Teachers – The Feedback Loop

- **The 5-Minute Rule**: If an assignment takes longer than expected, trim the length, not the rigor.

- **Closing the Loop**: Review 1–2 problems in class next day to validate effort and address common errors.

- **Differentiation Tip**: Allow students to choose 3 out of 5 tasks per night to encourage autonomy while ensuring practice across skills.

Complete Output Delivered: A structured, teacher-ready homework menu for Grade 8 Linear Equations.

Prompt: Educational Game Generator

Simple Prompt: "Act as an 'Gamification Coach' and 'Teacher Support Specialist'. Your task is to create a teacher-ready menu of educational games for [SUBJECT/TOPIC] designed for [GRADE LEVEL/COURSE]".

- **Tone**: Playful, Engaging
- **Format**: Game Plan
- **Platform**: Classroom Activities, Workshops
- **AI Role**: Gamification Coach/Teacher Support Specialist
- **Prompt Goal**: Engage students through playful learning experiences
- **Tags**: #games, #gamification, #engagement
- **Prompt Variables**: {topic}, {grade/level}, {gameFormat}

Best Practice Tip: Test games with a small group before scaling to the whole class.

Enhanced Prompt:

Act as an 'Gamification Coach' and 'Teacher Support Specialist'. Your task is to create a comprehensive menu of educational games for [SUBJECT/TOPIC] designed for [GRADE LEVEL/COURSE]. The goal is to generate game ideas that are not only highly engaging but are also strategically designed to reinforce core concepts, facilitate collaborative problem-solving, and provide authentic, low-stakes practice, thereby deepening conceptual understanding and retention.

The guide must be structured as follows:

1. **The "Why" it Matters**: A brief, impactful explanation of the core principle behind game-based learning (e.g., "Well-designed educational games create a 'flow state' of intense engagement and focus. They leverage intrinsic motivation through challenge, curiosity, and play, providing a powerful vehicle for practicing skills, applying knowledge, and receiving immediate feedback in a memorable and enjoyable way.").

2. **Overarching Learning Objectives**: List the 2-3 primary knowledge and skill standards these games are designed to reinforce and assess (e.g., "Knowledge: Recall key vocabulary and facts about [TOPIC]. Skill: Apply strategic thinking to solve problems related to [TOPIC]. Disposition: Collaborate effectively with peers to achieve a common goal.").

3. **The Game Menu**: A Strategic Mix Provide a list of 5-7 distinct game ideas. Ensure there is a deliberate mix of the following types to serve different classroom needs and resources:

    ◇ **Review & Drill Games**: For rapid recall and memorization.

    ◇ **Strategy & Problem-Solving Games**: For applying knowledge in complex, dynamic scenarios.

    ◇ **Collaborative & Simulation Games**: For fostering teamwork and understanding systems.

    ◇ **Creative & Narrative Games**: For demonstrating understanding through storytelling and creation.

4. **Breakdown for Each Game Idea**: For every game,

include:

- ◇ **Game Title**: A catchy, descriptive name.

- ◇ **Core Mechanics**: The basic actions of the game (e.g., "drawing cards," "rolling dice," "moving to-kens," "team trivia," "physical simulation").

- ◇ **Learning Objective**: The specific concept or skill the game targets.

- ◇ **How to Play**: A concise, step-by-step explanation of the rules and win condition.

- ◇ **Materials Needed**: A simple list of required items (e.g., 'index cards,' 'whiteboards,' 'a single die,' 'projector').

- ◇ **The "Why" it's Important**: A brief explanation of the cognitive or social skill it develops (e.g., "This game transforms rote memorization into an exciting challenge, significantly increasing retention through spaced repetition and active recall.").

5 **The "Why" Behind This Strategy**: A concluding summary explaining how this curated menu of games supports differentiated instruction, provides formative assessment data in a non-threatening way, and builds a positive classroom culture where learning is associated with joy and curiosity.

6 **Bonus**: Pro-Tip for Teachers - Facilitating Game-Based Learning:

- ◇ **Frame it as a Challenge**: Introduce games as "missions" or "challenges," not just fun activities, to maintain academic rigor.

- ◇ **Manage the Energy**: Have a clear signal for

pausing the game (c.g., a lights-off cue) for giving instructions or debriefing.

◇ **The Debrief is Key**: Always allocate 5 minutes after the game to ask, "What strategy worked? What did you learn? How does this connect to what we're studying?" This solidifies the learning objective.

◇ **Keep it Simple**: The best games have easy-to-learn rules. Complexity should be in the thinking, not the instructions.

The final output should be a complete, plug-and-play game kit that provides clear value, is easy to set up, and transforms the classroom into a dynamic learning environment.

Complete Output:

The "Why" it Matters: Games are not a diversion from learning; they are a powerful medium for it. They lower the affective filter, reducing the fear of failure. When students are playing, they are more willing to take risks, think critically, and persevere through challenges. This leads to deeper, more durable learning and a classroom culture that celebrates inquiry and effort.

The Game Menu: Learning Through Play

1. Game: Concept Charades

- **Core Mechanics**: Physical simulation, team guessing.
- **Learning Objective**: To review and recall key vocabulary and concepts from [TOPIC].
- **How to Play**: Divide the class into two teams. A player from one team draws a term or concept from a hat and must act it out without speaking. Their team has 60 seconds to guess. Points are awarded for correct guesses.

- **Materials Needed**: Slips of paper with key terms, a timer, a hat or bowl.
- **Why It's Important**: This game forces students to deeply understand a concept to its core in order to physicalize it, moving beyond definitional knowledge to conceptual mastery. It's also a fantastic energy booster.

2. Game: Review Relay

- **Core Mechanics**: Team competition, quick response.
- **Learning Objective**: To quickly retrieve factual information.
- **How to Play**: Teams line up. The teacher asks a question. The first student in each line races to the whiteboard to write the answer. The first correct answer wins a point for their team. The line rotates after each question.
- **Materials Needed**: Whiteboard or smartboard, markers.
- **Why It's Important**: This gamifies formative assessment, providing immediate feedback on class-wide understanding. The physical movement increases oxygen flow to the brain, enhancing alertness and memory.

3. Game: Design a Card Game

- **Core Mechanics**: Creative construction, systems thinking.
- **Learning Objective**: To demonstrate understanding of the relationships between concepts in [TOPIC] (e.g., historical events, parts of an ecosystem, literary elements).
- **How to Play**: In small groups, students create a simple card game. They must design cards (e.g., "Revolutionary War" card) and define rules for how they interact (e.g., "This card 'trumps' that card because..."). Groups

then swap and play each other's games.

- **Materials Needed**: Index cards, markers.
- **Why It's Important**: This is a high-level application activity. To create the game, students must analyze how concepts connect, compare, and influence each other, demonstrating synthesis and creativity.

4. Game: The "What If?" Simulation

- **Core Mechanics**: Role-play, collaborative problem-solving.
- **Learning Objective**: To apply knowledge of [TOPIC] to a novel scenario and predict outcomes.
- **How to Play**: Present a "What if?" scenario related to your topic (e.g., "What if the protagonist made a different choice?" "What if this scientific discovery never happened?"). In groups, students role-play stakeholders and must collaborate to manage the consequences.
- **Materials Needed**: Scenario prompt, role cards (optional).
- **Why It's Important**: This builds critical thinking, empathy, and a flexible understanding of systems. It shows that knowledge is not static but a tool for navigating complex, unpredictable situations.

5. Game: Quiz-Qui-Trade

- **Core Mechanics**: Peer-to-peer interaction, constant review.
- **Learning Objective**: To practice and explain concepts through repeated verbal rehearsal.
- **How to Play**: Each student has a flashcard with a question and answer. They mix around the room, find a partner, and quiz each other. After both answer, they trade cards and find a new partner. The cycle repeats.

- **Materials Needed**: Pre-made flashcards (one per student).

- **Why It's Important**: This strategy combines movement, social learning, and the "protégé effect," where students solidify their own understanding by teaching their peers. It ensures 100% participation.

The "Why" Behind This Strategy: This plan works because it aligns fun with function. Each game is a stealth assessment tool that provides you with real-time data on student understanding while the students are simply "playing." This approach meets students' needs for social interaction, movement, and fun, thereby dramatically increasing engagement and making learning stick.

Bonus: Pro-Tip for Teachers - Facilitating Game-Based Learning:

- **Start with Structure**: Begin with highly structured games (like Relay or Quiz-Qui-Trade) to establish clear routines and expectations for game play.

- **Focus on Transfer**: After the game, always debrief by explicitly connecting the game's content back to the larger unit of study. Ask: "How did what you did in this game help you understand [TOPIC] better?"

- **Embrace the Noise**: Game-based learning is active and sometimes noisy. Define the difference between productive "learning noise" and disruptive noise for your students beforehand.

Prompt: Field Trip Idea Generator

Simple Prompt: "Act as an 'Experiential Learning Architect' and 'Field Trip Planner'. Your task is to create a teacher-ready menu of curriculum-aligned field trip ideas for [TOPIC/UNIT] designed for [GRADE LEVEL]".

- **Tone**: Creative, Practical
- **Format**: Field Trip Plan
- **Platform**: School Planning, Parent Notes
- **AI Role**: Experiential Learning Architect/Field Trip Planner
- **Prompt Goal**: Connect classroom lessons with real-world experiences
- **Tags**: #fieldTrips, #experientialLearning, #engagement
- **Prompt Variables**: {subject}, {grade/level}, {learningGoals}

Best Practice Tip: Provide virtual alternatives for schools with limited travel budgets.

Enhanced Prompt:

Act as an 'Experiential Learning Architect' and 'Field Trip Planner'. Your task is to create a comprehensive menu of field trip ideas for [TOPIC/UNIT] designed for [GRADE LEVEL]. The goal is to generate ideas that are not just engaging outings but are powerful, direct extensions of the curriculum, designed to make abstract concepts tangible, foster curiosity, and create lasting memories that anchor classroom learning in the real world.

The guide must be structured as follows:

1. **The "Why" it Matters**: A brief, impactful explanation of the core principle behind experiential learning (e.g., "A field trip is the ultimate form of authentic assessment and engagement. It transforms passive learners into active investigators, allowing them to see, touch, and experience the concepts they study. This creates powerful 'episodic memories' that dramatically enhance retention, contextualize knowledge, and inspire future academic and personal interests.").

2. **Overarching Learning Objectives**: State the 2-3 primary curricular goals these trips are designed to achieve (e.g., "To connect classroom theory on [TOPIC] to real-world applications," "To conduct firsthand observation and data collection," "To interact with professionals and understand career pathways related to [TOPIC].").

3. **The Field Trip Menu**: A Spectrum of Experiences Provide a list of 5-7 distinct field trip ideas. Ensure there is a deliberate mix of the following types to cater to different learning styles and logistical possibilities:

   ◇ **Museum & Cultural Institution**: For curated, object-based learning.

   ◇ **Working Facility & Behind-the-Scenes Tour**: For seeing processes, STEM applications, and careers in action.

   ◇ **Natural Environment & Outdoor Site**: For observation, data collection, and environmental science.

   ◇ **Civic & Community Institution**: For understanding government, society, and community roles.

◇ **Performance & Artistic Venue**: For experiencing cultural expression and historical narrative.

4 **Breakdown for Each Field Trip Idea**: For every idea, include:

◇ **Destination Name & Type**: A specific example (e.g., "A Local Water Treatment Plant (Working Facility)").

◇ **Direct Curriculum Link**: How the trip directly reinforces the specific topic of [TOPIC] (e.g., "This trip brings our chemistry unit on purification and states of matter to life, showing large-scale application of filtration and distillation.").

◇ **The "Why" it's Important**: A brief explanation of the long-term impact and skills gained (e.g., "This demystifies a essential civic utility, fosters scientific literacy, and exposes students to potential engineering careers they may never have considered.").

◇ **Pre-Trip Activity Idea**: A small task to prepare students (e.g., "Generate questions for the plant manager about challenges and processes.").

◇ **On-Site Task**: A specific "mission" or focus for students (e.g., "Sketch and label the water purification process flow chart.").

5 **The "Why" Behind This Strategy**: A concluding summary explaining how this intentional approach to field trips—with pre-learning, on-site tasks, and post-reflection—moves beyond a simple day out and becomes a cornerstone of the unit, fostering deeper inquiry, relevance, and a tangible connection between academic work and the wider world.

6   **Bonus**: Pro-Tip for Teachers - Maximizing Impact:

◇ **The Holy Trinity of Field Trips**: A successful trip has three parts: 1) Pre-Trip Preparation (build background knowledge and curiosity), 2) On-Site Task (give students a focused job), and 3) Post-Trip Debrief (synthesize learning and share findings).

◇ **Virtual Option**: Always note if a high-quality virtual tour or digital experience is available for the destination as a backup or extension.

◇ **Community Connections**: Don't overlook small local businesses, parks, or civic centers. A trip to a local bakery can be a powerful lesson in chemistry, math, and entrepreneurship.

◇ **Chaperone Briefing**: Prepare chaperones with a guide on the learning objectives and key questions to ask students to keep the focus on learning.

The final output should be a complete, plug-and-play experiential learning guide that provides a clear rationale, curriculum connection, and practical steps for execution, transforming a field trip from a reward into an essential instructional strategy.

Complete Output:

The "Why" it Matters: Field trips are not a break from learning; they are learning at its most potent. They provide the sensory-rich experiences that form the foundation for abstract thought and critical thinking. By stepping into a real-world context, students understand the "why" behind their studies, see career possibilities, and develop a sense of place and connection to their community that cannot be replicated in the classroom.

The Field Trip Menu: Connecting [TOPIC] to the Real World

1. Destination: A Local History Museum Exhibit on [Relevant Era/Topic]

- **Type**: Museum & Cultural Institution
- **Curriculum Link**: Makes historical figures, events, and cultural artifacts from our unit physically present and real, moving beyond textbooks and images.
- **Long-Term Impact**: Fosters a tangible connection to the past, develops primary source analysis skills, and can spark a lifelong interest in history and preservation.
- **Pre-Trip Activity**: Students choose one artifact they hope to see and research its background.
- **On-Site Task**: Students complete a "scavenger hunt" to find specific artifacts and explain their significance.

2. Destination: A State or National Park with a Distinct Ecosystem

- **Type**: Natural Environment & Outdoor Site
- **Curriculum Link**: Provides a living lab for our unit on ecology, geology, or biodiversity, allowing for firsthand observation and data collection.
- **Long-Term Impact**: Instills an appreciation for conservation and the environment, teaches field research methods, and connects physical health and well-being to learning.
- **Pre-Trip Activity**: Learn to identify 5 key native plant or animal species.
- **On-Site Task**: In small groups, students conduct a transect study or sketch a detailed landscape, noting biotic and abiotic factors.

3. Destination: A Courtroom (Observing Proceedings)

- **Type**: Civic & Community Institution
- **Curriculum Link**: Brings our government and civics unit to life, illustrating the judicial branch in action, from courtroom procedure to the roles of officers of the court.
- **Long-Term Impact**: Demystifies the justice system, reinforces the importance of rule of law, and introduces powerful civic and career role models (judges, lawyers, clerks).
- **Pre-Trip Activity**: Study the roles of the judge, prosecutor, defense attorney, and bailiff. Review basic courtroom etiquette.
- **On-Site Task**: Students outline the process of a case they observe and identify the constitutional rights in play.

4. Destination: A Performing Arts Center (Rehearsal or Matinee)

- **Type**: Performance & Artistic Venue
- **Curriculum Link**: Connects to units on literature, drama, history, or cultural studies by showing the interpretive process of adapting text to performance.
- **Long-Term Impact**: Builds cultural capital, fosters empathy by seeing different perspectives acted out, and illustrates the collaborative nature of the arts.
- **Pre-Trip Activity**: Read the play or story being performed and discuss the themes.
- **On-Site Task**: Students note one directorial choice (e.g., lighting, costume, set design) and analyze how it influenced the story's mood or message.

5. Destination: Maker Space or Fabrication Lab (Fab Lab)

- **Type**: Working Facility & Behind-the-Scenes Tour

- **Curriculum Link**: Provides a physical application for STEM concepts (physics, engineering, design, math), showcasing the iteration process from design to proto-type.

- **Long-Term Impact**: Promotes a growth mindset through hands-on, trial-and-error learning, exposes students to advanced manufacturing technology, and fosters innovation and creativity.

- **Pre-Trip Activity**: Brainstorm a simple problem that could be solved with a designed object.

- **On-Site Task**: Students receive a brief tutorial on one tool (e.g., 3D printer, laser cutter) and sketch a design for a simple object.

The "Why" Behind This Strategy: This plan works because it treats field trips as integral, not incidental. By meticulously aligning each destination with curriculum objectives and structuring activities around preparation, engagement, and reflection, we ensure the experience is a formative assessment of understanding, not just a formative day of fun. This strategic approach justifies the investment of time and resources by demonstrating clear, impactful learning outcomes.

Bonus: Pro-Tip for Teachers - Maximizing Impact:

- **The Debrief is Everything**: Learning solidifies when you return. Students must create a presentation, write a thank-you letter to the host that explains what they learned, or create a museum-style exhibit for the school hallway based on their trip.

- **Chaperone Guide**: Give chaperones a cheat sheet with 3-5 questions to ask their group (e.g., "What do you think that machine is for?" "Why do you think the artist made that choice?"). This turns them into learning facilitators

Prompt: Technology Integration Ideas

Simple Prompt: "Act as an 'Educational Technology Integration Coach'. Your task is to create a teacher-ready guide for integrating technology into classroom learning into the classroom for [SUBJECT/TOPIC] and [GRADE LEVEL]".

- **Tone**: Innovative, Practical
- **Format**: Idea List
- **Platform**: Classroom, Edtech Planning
- **AI Role**: Educational Technology Integration Coach
- **Prompt Goal**: Suggest tools and strategies for effective tech use
- **Tags**: #edtech, #technology, #teaching
- **Prompt Variables**: {topic}, {tools}, {grade/level}

Best Practice Tip: Always ensure tech tools align with accessibility standards and student needs.

Enhanced Prompt:

Act as an 'Educational Technology Integration Coach'. Your task is to create a comprehensive guide to integrating technology into the classroom for [SUBJECT/TOPIC] and [GRADE LEVEL]. The goal is to provide a toolkit of ideas that move beyond using tech as a mere substitution for paper-and-pencil tasks, and instead, leverage it to transform learning by enabling creation, fostering collaboration, and providing personalized pathways that were previously impossible.

The guide must be structured as follows:

1 **The "Why" it Matters**: A brief, impactful explanation of the core principle behind effective tech integration (e.g., "Technology is not a subject to be taught; it is a vehicle for deeper learning. The goal of integration is not to use more tech, but to use it purposefully to amplify student voice, transform the learning experience, and accelerate the achievement of curricular goals. The best tech integration is invisible—it's simply the best tool for the learning job.").

2 **Overarching Learning Objectives**: State the 2-3 primary skills this tech integration is designed to enhance (e.g., "To foster collaborative problem-solving," "To provide authentic avenues for creative expression," "To offer differentiated, immediate feedback and practice.").

3 **The Integration Menu**: A Strategic Mix Provide a list of 5-7 distinct integration ideas. The ideas must be categorized using the SAMR Model framework to show progression from enhancement to transformation:

    ◇ **Substitution/Augmentation**: Tech acts as a direct tool substitute with functional improvement (e.g., word processor instead of paper).

    ◇ **Modification/Redefinition**: Tech allows for significant task redesign and the creation of new, previously inconceivable tasks (e.g., collaborating globally on a document).

4 **Breakdown for Each Integration Idea**: For every idea, include:

    ◇ **Idea Name & Tech Tool**: A specific tool or platform (e.g., "Digital Storytelling with Adobe Spark Video").

    ◇ **SAMR Level**: Label it as Substitution/Augmentation or Modification/Redefinition.

◇ **Core Task**: A 1-2 sentence description of the student learning activity.

◇ **The "Why" it's Important**: A brief explanation of the specific pedagogical benefit and long-term impact (e.g., "This transforms student from consumers of information into creators of knowledge, building digital literacy, narrative skills, and confidence in a shareable format.").

◇ **Ideal Use Case**: The specific learning context where this idea shines.

5 **The "Why" Behind This Strategy**: A concluding summary explaining how this purposeful, tiered approach to technology prevents "tech for tech's sake," ensures alignment with learning goals, and prepares students with the critical digital citizenship and literacy skills they need for the future.

6 **Bonus**: Pro-Tip for Teachers - The Integration Framework:

◇ **Start with the Learning Goal, Not the App**: Never ask "What cool thing can I do with this app?" Instead, ask "What is my learning objective?" and then find the tech that best helps students achieve it.

◇ **The SAMR Ladder**: Don't feel pressure to be at "Redefinition" for every lesson. A healthy mix across the SAMR model is perfect. Sometimes, Augmentation (like using a quiz game for review) is exactly what's needed.

◇ **Always Have Plan B**: Technology will fail. Always have a low-tech or no-tech version of the activity ready to go.

◇ **Model Digital Citizenship**: Use these tools to explicitly teach and model respectful online communication, copyright, and digital footprints.

The final output should be a complete, plug-and-play tech integration guide that empowers educators to use technology as a powerful lever for student engagement and mastery.

Complete Output:

The "Why" it Matters: True technology integration is when the use of the tool becomes seamless and essential to the learning goal. It's not about gamifying to make learning "fun"; it's about empowering students to do real-world work, connect with authentic audiences, and demonstrate their understanding in ways that were never before possible. This shifts the student role from passive recipient to active creator and collaborator.

The Integration Menu: Purposeful Tech for Learning

1. Idea: Collaborative Brainstorming & Inquiry with Padlet

- **SAMR Level**: Modification
- **Core Task**: Instead of a KWL chart on chart paper, students use Padlet (a digital corkboard) to post questions, ideas, images, and links at the start of a unit. They can comment on and "upvote" each other's posts.
- **Why it's Important**: It captures the whole class's prior knowledge and curiosity in a dynamic, interactive, and visually engaging way. It fosters a community of inquiries and provides the teacher with a map of student interests and misconceptions to guide instruction.
- **Ideal Use Case**: Launching a new unit, generating research questions, or a unit-ending "what we learned" reflection.

2. Idea: Formative Assessment with Kahoot! or Quizizz

- **SAMR Level**: Augmentation
- **Core Task**: Use a game-based platform to conduct a quick, fun review or check for understanding. Students answer on their devices, and results are displayed in real-time.
- **Why It's Important**: It provides instant, whole-class data on student understanding in an highly engaging format. The game mechanics increase energy and participation, and immediate feedback helps students identify gaps in their knowledge.
- **Ideal Use Case**: Mid-unit review, vocabulary practice, or a pre-test to gauge baseline knowledge.

3. Idea: Digital Portfolios with Google Sites or SeeSaw

- **SAMR Level**: Redefinition
- **Core Task**: Students create a personal website or digital journal where they upload photos of work, record videos explaining their thinking, and write reflections on their learning journey over time.
- **Why It's Important**: This transforms assessment from a one-time event to an ongoing process. It fosters metacognition, pride in work, and provides a rich, holistic picture of student growth for teachers and parents. It also teaches valuable website design and curation skills.
- **Ideal Use Case**: A year-long project to track growth in any subject, especially writing, art, or science.

4. Idea: Virtual Field Trips with Google Earth or 360° Videos

- **SAMR Level**: Modification/Redefinition
- **Core Task**: Instead of just reading about the Amazon rainforest or the Colosseum, students take an immer-

sive, guided tour using Google Earth's Voyager stories or curated YouTube 360° videos.

- **Why It's Important**: It breaks down the walls of the classroom, providing experiential learning that builds context and schema in a way photos in a textbook never could. It makes abstract concepts tangible and sparks curiosity.

- **Ideal Use Case**: Social studies units on geography and history, science units on ecosystems, literature settings.

5. Idea: "Explain Your Thinking" Videos with Flip (formerly Flipgrid)

- **SAMR Level**: Modification

- **Core Task**: Students record short videos of themselves solving a math problem, explaining a science concept, or analyzing a quote from a book, then watch and respond to their peers' videos.

- **Why It's Important**: It provides a window into student thinking processes that written work alone cannot. It builds oral communication skills and creates a supportive community of learners where students can teach each other.

- **Ideal Use Case**: Math problem-solving, reading comprehension checks, practicing world language speaking skills.

6. Idea: Choose-Your-Own-Adventure Stories with Google Slides

- **SAMR Level**: Redefinition

- **Core Task**: Students create interactive narratives where the reader makes choices that change the story's path. This is done by linking different slides together.

- **Why It's Important**: This requires deep understanding of narrative structure, cause and effect, and sequencing. It is a highly creative and engaging way to demonstrate comprehension and writing skills, moving far beyond a traditional essay.

- **Ideal Use Case**: Culminating project for literature or creative writing unit.

The "Why" Behind This Strategy: This plan works because it aligns technology with sound pedagogy. Each idea is not just a "cool tool"; it's a strategic choice to enhance a specific learning outcome. By providing a mix of low-threshold and high-ceiling ideas, it makes tech integration accessible for beginners and inspiring for experts, ensuring that technology truly serves the learning, not the other way around.

Bonus: Pro-Tip for Teachers - The Integration Framework:

- **The One-Thing Rule**: Don't try to master everything at once. Pick one new tool or strategy per semester and become an expert at it.

- **Student Tech Support**: Identify your "tech-savvy" students and empower them to be the first line of support for their classmates. This builds leadership and saves you time.

- **Focus on the Task, Not the Tool**: When giving instructions, focus on the learning objective (e.g., "persuade your audience") rather than the tool's features (e.g., "click this button").

Category 2: Student Engagement & Activities

Theme

This category is designed to boost student involvement by introducing a variety of structured, interactive activities. These activities include icebreakers to help students connect, debates to encourage critical thinking, role-plays that allow learners to explore different perspectives, and scavenger hunts that foster teamwork and exploration. Each strategy is grounded in evidence-based practices, with the primary goal of sparking curiosity and ensuring that students actively participate in their learning experiences.

Category Goal

The primary objective of providing teachers with these prompts is to offer practical, engaging tools that can be readily implemented in the classroom. These prompts are designed to motivate students, foster collaboration among peers, and create a learning environment that is both enjoyable and memorable. By utilizing these prompts, teachers can enhance student involvement, support meaningful interactions, and contribute to a dynamic classroom atmosphere where learning is both effective and fun.

Mini-Index: Student Engagement & Activities

Prompt	AI Role	Prompt Goal
Discussion Question Generator	Discussion Facilitator / Literature Coach	Spark critical thinking and participation
Debate Topic Generator	Debate Coach / Critical Thinking Facilitator	Provide structured, thought-provoking debate topics
Group Activity Generator	Collaborative Learning Coach / Classroom Engagement Specialist	Design group tasks that foster teamwork
Role-Play Scenario Script	Experiential Learning Coach / SEL Role-Play Specialist	Reinforce learning through immersive role-play
Icebreaker Activity Generator	Classroom Community Builder / Student Engagement Specialist	Build rapport and encourage participation
Classroom Scavenger Hunt Ideas	Classroom Gamification Coach / Experiential Learning Specialist	Combine fun and learning with interactive hunts
Classroom Bulletin Board Ideas	Learning Environment Designer / Classroom Visual Learning Specialist	Create visual displays that reinforce learning
Educational App Generator	EdTech Strategist / Digital Learning Designer	Inspire new app concepts for teaching topics
Classroom Decoration Ideas	Learning Environment Designer / Classroom Space Architect	Design engaging environments that support learning
Classroom Reward System	Behavior & Motivation Architect	Motivate students with fair, inclusive rewards

Detailed Prompts

Prompt: Discussion Question Generator

Simple Prompt: "Act as a 'Discussion Facilitator' and 'Literature Coach'. Your task is to create a structured discussion guide with sequenced questions for [BOOK/ARTICLE] tailored to a [GRADE LEVEL/COURSE] classroom".

- **Tone**: Engaging, Thought-Provoking
- **Format**: Question List (10)
- **Platform**: Classroom, Study Groups, Online Forums
- **AI Role**: Discussion Facilitator / Literature Coach
- **Prompt Goal**: Encourage deeper thinking and student participation
- **Tags**: #discussion, #studentEngagement, #criticalThinking
- **Prompt Variables**: {topic}, {grade/level}, {learningObjectives}

Best Practice Tip: Use a mix of open-ended and probing questions to keep the discussion dynamic and inclusive.

Enhanced Prompt:

Act as a 'Discussion Facilitator' and 'Literature Coach'. Your task is to create a comprehensive discussion guide for [BOOK/ARTICLE] tailored to a [GRADE LEVEL/COURSE] classroom. The goal is to generate questions that move beyond basic comprehension to foster critical thinking, textual analysis, and deep, student-led engagement with the themes and ideas presented.

The guide must be structured as follows:

1. **The "Why" it Matters**: A brief, impactful explanation of the core principle behind using leveled questioning to facilitate discussion (e.g., "A truly transformative discussion is scaffolded through a progression of questions, moving students from understanding the text to interrogating its meaning and connecting it to the wider world.").

2. **Overarching Discussion Goal**: State the primary learning objective for this discussion (e.g., "Students will be able to analyze how the author uses symbolism to critique social norms and engage in a Socratic seminar to defend their interpretations with textual evidence.").

3. **The "Questions" List**: Provide 10-15 carefully sequenced questions. The sequence should intentionally build through the following levels:

    ◇ **Stage 1**: Foundational (Comprehension & Recall): Questions that establish a common understanding of the text's key events and details.

    ◇ **Stage 2**: Analytical (Application & Analysis): Questions that require students to examine authorial choices, literary devices, structure, and character motivation.

    ◇ **Stage 3**: Generative (Synthesis & Evaluation): Questions that push students to connect the text to broader themes, other texts, and their own lives, evaluating its significance and arguing their perspectives.

4. **Breakdown for Each Question**: For every question, include:

    ◇ **The Question**: Phrased precisely as the teacher

should ask it.

◇ **Question Level**: Tag it as Foundational, Analytical or Generative.

◇ **Purpose & "Why" it's Important**: A brief explanation of the specific cognitive skill or literary element this question targets (e.g., "This question forces students to distinguish between a character's actions and their motivations, a key skill in character analysis.").

◇ **Facilitation Notes & Key Discussion Points**: Not just points to cover, but practical advice for the teacher. This should include:

  ▪ **Anticipated Student Responses**: What to listen for.

  ▪ **Potential Follow-Up Questions**: How to deepen the conversation if it stalls.

  ▪ **Differentiation**: How to simplify the question for struggling learners or extend it for advanced ones (e.g., "For students needing support, rephrase as a multiple choice. Challenge others to find a second theme the symbol represents.").

5 **The "Why" Behind This Strategy**: A concluding summary explaining how this structured, leveled approach builds student confidence, ensures all voices can participate at different levels, and creates a more rigorous and inclusive dialogue than a set of random questions.

6 **Bonus**: Pro-Tip for Facilitation: Add a short section on actionable facilitation strategies for this specific discus-

sion (e.g., "Begin with a silent 'Think-Ink' period where students jot down answers to the first two foundational questions to ensure everyone is prepared," "Use a 'talking stick' or structured protocol to ensure equitable airtime," "Conclude by asking students to reflect on how their understanding of the text changed through the discussion.").

The final output should be a complete, plug-and-play guide that empowers a teacher to facilitate a dynamic, student-centered discussion with confidence.

Complete Output

Sample Discussion Guide: To Kill a Mockingbird (Grade 9)

1. The "Why" it Matters

A rich discussion moves students beyond recalling plot details. By progressing from foundational to analytical to generative questions, students practice critical thinking, empathy, and evidence-based reasoning—skills that matter far beyond this novel.

2. Overarching Discussion Goal

Objective: Students will analyze how Harper Lee uses character development and symbolism to critique social norms and engage in a structured seminar to defend interpretations with textual evidence.

3. The Question List

Stage 1: Foundational (Comprehension & Recall)

- **Question**: Who are the main characters introduced in

the first chapters, and how does Scout describe them?

- ◇ **Level**: Foundational

- ◇ **Purpose**: Builds shared knowledge of character introductions.

- ◇ **Facilitation Notes**: Expect recall (Atticus, Jem, Scout, Dill).

- ◇ **Follow-Up**: "What words does Scout use that give us clues about her perspective?"

- ◇ **Differentiation**: Offer multiple-choice for struggling students; challenge others to cite direct quotations.

- **Question**: What events lead up to the children's fascination with Boo Radley?

- ◇ **Level**: Foundational

- ◇ **Purpose**: Ensures chronological understanding.

- ◇ **Facilitation Notes**: Look for mentions of myths, dares, and neighborhood rumors.

Stage 2: Analytical (Application & Analysis)

- **Question**: Why does Harper Lee choose Scout as the narrator? How does her perspective shape the story?

- ◇ **Level**: Analytical

- ◇ **Purpose**: Examines point of view as a literary device.

- ◇ **Facilitation Notes**: Students may note innocence, honesty, limited perspective.

- ◇ **Follow-Up**: "How would the story change if Atticus narrated?"

- **Question**: In what ways does Atticus serve as a symbol in the novel?

    ◇ **Level**: Analytical

    ◇ **Purpose**: Pushes symbolic interpretation.

    ◇ **Facilitation Notes**: Watch for connections to fairness, justice, moral courage.

    ◇ **Differentiation**: Scaffold with "What qualities does Atticus show in defending Tom Robinson?"

- **Question**: How does Harper Lee use setting (Maycomb) to reinforce themes of tradition and prejudice?

    ◇ **Level**: Analytical

    ◇ **Purpose**: Examines setting as a theme driver.

Stage 3: Generative (Synthesis & Evaluation)

- **Question**: What lessons about empathy and justice are timeless in this novel? How do they apply today?

    ◇ **Level**: Generative

    ◇ **Purpose**: Connects text to broader themes and real-world contexts.

    ◇ **Facilitation Notes**: Encourage connections to modern issues of inequality.

- **Question**: If you were rewriting this story today, what social issues would you highlight, and how?

    ◇ **Level**: Generative

    ◇ **Purpose**: Promotes creativity and synthesis.

    ◇ **Differentiation**: For advanced students, require parallels to current literature or media.

- **Question**: Do you think Atticus is portrayed as too idealized? Why or why not?

  ◇ **Level**: Generative

  ◇ **Purpose**: Encourages evaluation of character complexity.

  ◇ **Facilitation Notes**: Look for debate; push students to justify with text.

4. The "Why" Behind This Strategy

This level questioning ensures all students can engage: those needing confidence begin with recall, while advanced thinkers can synthesize and critique. By sequencing, we avoid shallow discussion and instead build toward a richer, student-led conversation.

5. Bonus: Pro-Tip for Facilitation

- **Silent Start**: Begin with a 3-minute "Think-Write" on Question 1 to ensure everyone has an entry point.
- **Equity of Voice**: Use a talking stick or digital protocol (e.g., each student posts one idea on Padlet before speaking).
- **Closing Reflection**: End by asking: "What's one idea you hadn't considered before this discussion?"

This guide is now a ready-to-use Complete Output, with sequenced questions, facilitation notes, and strategies built-in.

Prompt: Debate Topic Generator

Simple Prompt: "Act as an 'Debate Coach' and 'Critical Thinking Facilitator'. Your task is to create a structured menu of compelling, curriculum-aligned debate topics for [SUBJECT/COURSE] designed for [GRADE LEVEL]".

- **Tone**: Neutral, Stimulating
- **Format**: Debate Topic List
- **Platform**: Classroom, Debate Clubs
- **AI Role**: Debate Coach / Critical Thinking Facilitator
- **Prompt Goal**: Provide structured debate topics that spark critical engagement
- **Tags**: #debate, #studentEngagement, #discussion
- **Prompt Variables**: {subject/topic}, {grade/level}

Best Practice Tip: Allow students to pick sides anonymously to encourage honest participation and reduce bias.

Enhanced Prompt:

Act as a 'Debate Coach' and 'Critical Thinking Facilitator'. Your task is to create a comprehensive menu of compelling debate topics for [SUBJECT/COURSE] designed for [GRADE LEVEL]. The goal is to generate topics that are not only engaging but are also strategically designed to be balanced, researchable, and capable of fostering rigorous critical thinking, empathetic listening, and evidence-based argumentation in students.

The guide must be structured as follows:

1 **The "Why" it Matters**: A brief, impactful explanation of the core principle behind academic debate (e.g., "De-

bate is the ultimate exercise in critical thinking. It requires students to deeply understand multiple perspectives, evaluate the strength of evidence, construct logical arguments, and respond dynamically to counter-claims. This process builds intellectual humility, communication skills, and the ability to engage in civil discourse—cornerstones of informed citizenship and academic excellence.").

2 **Overarching Learning Objectives**: List the 2-3 primary cognitive and social-emotional skills this debate unit is designed to develop (e.g., "Skill: Construct a coherent argument using credible evidence. Skill: Anticipate and effectively rebut counterarguments. Disposition: Engage in respectful discourse with those who hold opposing viewpoints.").

3 **The Debate Menu**: A Spectrum of Deliberation Provide a list of 5-7 distinct debate topics. Ensure there is a deliberate mix of the following types to cater to diverse interests and align with different facets of the subject:

   ◇ **Policy Debate**: Topics centered on proposing a change to a current law, rule, or policy.

   ◇ **Value Debate**: Topics that grapple with ethical dilemmas, philosophical questions, and conflicts in core values.

   ◇ **Fact-Based Debate**: Topics that center on the interpretation of data, historical events, or scientific findings.

   ◇ **Comparative Advantage**: Topics that debate the relative merits of two different ideas, systems, or approaches.

4 **Breakdown for Each Debate Topic**: For every topic, include:

⬦ **The Resolution**: Phrased as a clear, neutral, and debatable statement (e.g., "Resolved: That social media platforms should be held legally responsible for content posted by their users.").

⬦ **Debate Type**: Categorize it from the list above (e.g., "Policy Debate").

⬦ **The "Why" it's Important**: A brief explanation of the topic's relevance and the long-term skills it develops (e.g., "This topic forces students to grapple with complex issues of free speech, corporate responsibility, and legal precedent, connecting classroom learning to real-world digital citizenship.").

⬦ **Key Argument Starters (For Affirmative & Negative)**: 2-3 potential core arguments for each side to help students and teachers begin their research (e.g., "Affirmative: Duty of care, amplifying harmful rhetoric. Negative: Chilling effect on free expression, impracticality of moderation.").

5 **The "Why" Behind This Strategy**: A concluding summary explaining how this menu of balanced, well-framed topics promotes deeper intellectual engagement than simple discussion, teaches students to argue with ideas instead of emotions, and creates a classroom environment where critical thinking is visibly practiced and valued.

6 **Bonus**: Pro-Tip for Teachers - Facilitating productive Debate: Add a short section with actionable advice for implementation.

⬦ **Format Suggestion**: "For beginners, use a structured format like 'Lincoln-Douglas' or a simple 'Pro/Con' fishbowl debate to ensure clarity and order."

◇ **Research & Preparation**: "Require students to prepare a brief 'casebook' with evidence for both sides of the resolution. This ensures they understand the totality of the issue and are not just defending a pre-assigned position blindly."

◇ **Setting Norms**: "Co-create and enforce strict norms of respectful engagement before beginning. Use a talking piece or timed speeches to ensure equitable participation."

◇ **The Role of the Audience**: "Assign the audience specific tasks, such as tracking arguments on a T-chart or serving as 'cross-examiners,' to keep all students actively engaged."

The final output should be a complete, plug-and-play debate module that provides a clear path from topic selection to structured argumentation, empowering students to engage in respectful, evidence-based, and critical discourse.

Complete Output

Grade Level: 10 | Subject: Social Studies | Topic: Debate

1. The "Why" it Matters

Debate is one of the most effective ways to sharpen critical thinking. It requires students to research multiple perspectives, build evidence-based claims, and practice respectful listening. In a world of polarization, teaching students to argue ideas—not people—is an essential civic skill.

2. Overarching Learning Objectives:

- **Cognitive Skill**: Construct logical, well-supported arguments using credible evidence.
- **Cognitive Skill**: Anticipate counterarguments and formulate respectful rebuttals.
- **Disposition**: Develop empathy and intellectual humility by engaging with opposing perspectives.

3. The Debate Menu: A Spectrum of Deliberation

1. Policy Debate

- **Resolution**: Resolved: That voting should be mandatory in national elections.
- **Why It's Important**: Encourages analysis of civic responsibility, individual rights, and global democratic practices.
- **Key Argument Starters**:
- **Affirmative**: Increases voter turnout, strengthens democracy, ensures diverse representation.
- **Negative**: Violates freedom of choice, risks uninformed voting, may increase resentment.

2. Value Debate

- **Resolution**: Resolved: Protecting free speech is more important than preventing misinformation.
- **Why It's Important**: Students wrestle with ethical tensions between liberty and responsibility in the digital age.
- **Key Argument Starters**:
- **Affirmative**: Free speech is foundational to democracy, censorship is dangerous.
- **Negative**: Misinformation undermines public trust,

harms society, and endangers lives.

3. Fact-Based Debate

- **Resolution**: Resolved: That climate change is the greatest global security threat of the 21st century.
- **Why It's Important**: Encourages examination of scientific data and its geopolitical implications.
- **Key Argument Starters**:
- **Affirmative**: Rising sea levels, resource wars, refugee crises.
- **Negative**: Other threats (nuclear proliferation, cyberwarfare, pandemics) are more immediate.

4. Comparative Advantage Debate

- **Resolution**: Resolved: That online learning provides more benefits than in-person learning.
- **Why It's Important**: Connects to students' own lived experiences and invites balanced, evidence-driven evaluation.
- **Key Argument Starters**:
- **Affirmative**: Flexible, accessible, cost-effective, personalized pacing.
- **Negative**: Loss of social skills, weaker community building, digital divide issues.

5. Policy Debate

- ◇ **Resolution**: Resolved: That the United Nations should have stronger authority to enforce international law.

- ◇ **Why It's Important**: Deepens understanding of global governance, sovereignty, and cooperation.

◇ **Key Argument Starters**:

◇ **Affirmative**: Improves peacekeeping, accountability, global stability.

◇ **Negative**: Violates national sovereignty, risks misuse of power, impractical enforcement.

4. The "Why" Behind This Strategy

This curated menu balances fact, value, policy, and comparative advantage debates. Students learn to analyze evidence, respectfully disagree, and practice civic discourse in a structured way. This approach moves beyond casual discussion to rigorous, skills-based engagement.

5. Bonus: Pro-Tip for Teachers – Facilitating Debate

- **Format Suggestion**: Start with a simple Pro/Con fishbowl debate before moving to formal styles like Lincoln-Douglas.

- **Research Requirement**: Have students prepare evidence for both sides before they know their assigned position.

- **Norm-Setting**: Co-create norms of respect (no interruptions, cite evidence, listen actively).

- **Audience Role**: Non-speakers should track arguments on a T-chart or serve as judges to keep engagement high.

This is a ready-to-use debate menu that a teacher can drop into a Social Studies unit, with structured resolutions, rationale, and scaffolds for both sides.

Prompt: Group Activity Generator

Simple Prompt: "Act as a 'Collaborative Learning Coach' and 'Classroom Engagement Specialist'. Your task is to create a structured menu of purposeful group activity ideas for the topic of [TOPIC] designed for [GRADE LEVEL/COURSE]".

- **Tone**: Collaborative, Practical
- **Format**: Group Activity Plan
- **Platform**: Classroom, Workshops
- **AI Role**: Collaborative Learning Coach / Classroom Engagement Specialist
- **Prompt Goal**: Facilitate teamwork and collective problem-solving
- **Tags**: #groupWork, #collaboration, #studentEngagement
- **Prompt Variables**: {topic}, {groupSize}, {grade/level}

Best Practice Tip: Assign rotating roles (e.g., leader, recorder, presenter) so every student stays engaged.

Enhanced Prompt:

Act as a 'Collaborative Learning Coach' and 'Classroom Engagement Specialist'. Your task is to create a comprehensive menu of structured group activity ideas for the topic of [TOPIC] designed for [GRADE LEVEL/COURSE]. The goal is to generate activities that are not only engaging but are also strategically designed to promote positive interdependence, individual accountability, and the development of crucial social and cognitive skills through purposeful collaboration.

The guide must be structured as follows:

1. **The "Why" it Matters**: A brief, impactful explanation of the core principle behind effective group work (e.g., "True collaborative learning moves beyond students simply sitting together. It is a high-leverage strategy that creates a 'sink or swim together' environment, forcing the negotiation of ideas, peer teaching, and the development of communication and leadership skills that are critical for future success in academia and the workplace.").

2. **Alignment to Learning Objectives**: State the 2-3 primary knowledge and social skills this set of activities is designed to reinforce (e.g., "Knowledge: Students will synthesize the causes of the American Revolution. Skills: Students will collaborate to create a consensus-based product; they will delegate tasks and manage time effectively.").

3. **The Activity Menu**: Structured for Success Provide a list of 5-7 distinct activity ideas. Ensure there is a deliberate mix of the following types to serve different purposes:

   ◇ **Problem-Based Learning (PBL)**: Solving a complex, real-world problem.

   ◇ **Creative Construction**: Building a physical or digital model, story, or artwork.

   ◇ **Peer Review & Analysis**: Critiquing, editing, or evaluating work against a set of criteria.

   ◇ **Jigsaw**: Becoming an "expert" on one subtopic to teach teammates.

   ◇ **Simulation & Role-Play**: Acting out a scenario to understand different perspectives.

4 **Breakdown for Each Activity Idea**: For every idea, include:

◇ **Activity Title**: A catchy, descriptive name.

◇ **Core Task**: A 1-2 sentence description of the group's ultimate goal.

◇ **Group Size & Roles**: Suggest an ideal group size & define specific, meaningful roles for each member to ensure individual accountability (e.g., "Facilitator: Keeps the group on task. Recorder: Documents key ideas. Materials Manager: Gathers supplies. Spokesperson: Presents to the class").

◇ **Timeframe**: A realistic time commitment for the activity (e.g., "30-45 minutes").

◇ **The "Why" it's Important**: A brief explanation of the specific collaborative and cognitive skills it targets (e.g., "This task requires consensus-building and practical application of knowledge, moving beyond memorization to deep synthesis and teamwork.").

◇ **Tangible Outcome**: The product the group must create (e.g., "A labeled diagram," "A 3-minute podcast summary," "A solution proposal on chart paper.").

5 **The "Why" Behind This Strategy**: A concluding summary explaining how this approach to group work—with clear roles, tasks, and outcomes—prevents common pitfalls like "social loafing," ensures all students are active participants, and creates a more dynamic and inclusive learning environment than individual or whole-class work alone.

6 **Bonus**: Pro-Tip for Teachers - Facilitating Effective Collaboration: Add a short section with actionable advice for implementation.

◇ **Setting Norms**: "Co-create group work norms with your class beforehand (e.g., 'Everyone contributes,' 'Disagree respectfully,' 'Use quiet voices')."

◇ **Formative Checks**: "Implement a 'group halt' midway through for a quick status check—'What is your biggest success so far? What is your biggest challenge?'"

◇ **Assessment**: "Use a simple group/self-assessment rubric post-activity. Have students rate their group on collaboration and themselves on their contribution to hold them accountable."

◇ **Group Formation**: "Strategically form groups to mix skill levels & strengths; avoid allowing students to always choose their own partners to maximize diversity of thought."

The final output should be a complete, plug-and-play collaborative learning kit that provides a clear framework for execution, ensures meaningful participation from all students, and transforms group work from a classroom management challenge into a powerful engine for learning.

Complete Output

Grade Level: 8 | Subject: Science | Topic: Ecosystem

1. The "Why" it Matters

True collaborative learning is not about dividing the work; it's about co-creating understanding. In ecosystems, just like in group work, balance and interdependence are essential. These activities are designed to help students practice teamwork, negotiation of ideas, and peer teaching, while deepening their grasp of how ecosystems function and change.

2. Alignment to Learning Objectives

- **Knowledge Objective**: Students will explain the flow of energy in ecosystems and the impact of human activity.
- **Skill Objective 1**: Students will work collaboratively to create consensus-driven products.
- **Skill Objective 2**: Students will practice communication, task delegation, and leadership in a group setting.

3. The Activity Menu: Structured for Success

1. "Ecosystem in Crisis" (Problem-Based Learning)

- ◇ **Core Task**: Groups design a recovery plan for a local ecosystem affected by pollution or invasive species.

- ◇ **Group Size & Roles**: 4–5 students (Facilitator, Recorder, Researcher, Materials Manager, Presenter).

- ◇ **Timeframe**: 45 minutes.

- ◇ **Why It's Important**: Builds problem-solving, consensus-building, and application of science concepts.

- ◇ **Tangible Outcome**: A poster or slide outlining the recovery plan.

2. "Food Web Builders" (Creative Construction)

◇ **Core Task**: Groups construct a large, labeled food web using string and cards representing organisms.

◇ **Group Size & Roles**: 3–4 students (Card Designer, Web Organizer, Connector, Explainer).

◇ **Timeframe**: 30 minutes.

◇ **Why It's Important**: Visualizes interdependence and energy flow while promoting teamwork.

◇ **Tangible Outcome**: A classroom displayable food web.

3. "Peer Scientists" (Peer Review & Analysis)

◇ **Core Task**: Groups swap mini-lab reports on decomposition and critique each other using rubric.

◇ **Group Size & Roles**: 3 students (Reviewer, Evidence Checker, Feedback Giver).

◇ **Timeframe**: 25 minutes.

◇ **Why It's Important**: Builds analytical skills and constructive peer feedback habits.

◇ **Tangible Outcome**: Annotated reports with peer comments.

4. "Jigsaw Ecosystem Experts" (Jigsaw Method)

◇ **Core Task**: Each student becomes an "expert" on one ecosystem (forest, desert, tundra, ocean) and teaches their group.

◇ **Group Size & Roles**: 4 students (Each = Expert Teacher).

◇ **Timeframe**: 40 minutes.

◇ **Why It's Important**: Encourages ownership of knowledge and peer teaching.

◇ **Tangible Outcome**: A group comparison chart across ecosystems.

5. "Council of Species" (Simulation & Role-Play)

◇ **Core Task**: Students role-play as species (e.g., wolf, rabbit, tree, decomposer) in a debate about a human policy decision (e.g., cutting down part of a forest).

◇ **Group Size & Roles**: 5–6 students (Each plays one role; 1 Moderator).

◇ **Timeframe**: 35 minutes.

◇ **Why It's Important**: Develops empathy, argumentation skills, and systems thinking.

◇ **Tangible Outcome**: A group "policy decision" with reasoning documented.

4. The "Why" Behind This Strategy

With defined roles, clear outcomes, and realistic timeframes, these activities reduce the risk of "social loafing" while increasing equity in participation. They make group work a dynamic engine of learning, not a management challenge. Students walk away with content knowledge, collaboration skills, and ownership of the process.

5. Bonus: Pro-Tip for Teachers – Facilitating Effective Collaboration

- **Setting Norms**: Co-create class group norms (e.g., "Everyone speaks at least once," "Respect turns," "Use evidence to justify ideas").

- **Formative Checks**: Pause mid-activity for a 2-minute "Group Pulse Check" (What's working? What's challenging?).

- **Assessment**: Use a simple rubric with two sections: Group Collaboration and Individual Contribution.

- **Group Formation**: Mix skill levels strategically; rotate roles weekly to prevent hierarchy.

This is now a plug-and-play activity kit for Grade 8 Science. Teachers can copy, adapt, and implement immediately.

Prompt: Role-Play Scenario Script

Simple Prompt: "Act as an 'Experiential Learning Coach' and 'SEL Role-Play Specialist'. Your task is to create a structured role-play scenario kit that immerses students in collaborative problem-solving and perspective-taking for [TOPIC] designed for [GRADE LEVEL/COURSE]".

- **Tone**: Creative, Immersive
- **Format**: Role-Play Script
- **Platform**: Classroom, Drama Workshops
- **AI Role**: Experiential Learning Coach / SEL Role-Play Specialist
- **Prompt Goal**: Reinforce concepts through interactive role-play
- **Tags**: #roleplay, #engagement, #experientialLearning
- **Prompt Variables**: {topic}, {roles}, {learningObjectives}

Best Practice Tip: Keep roles flexible so both introverted and extroverted students can contribute comfortably.

Enhanced Prompt

Act as an 'Experiential Learning Coach' and 'SEL Role-Play Specialist'. Your task is to create a comprehensive role-play scenario kit for [TOPIC] designed for [GRADE LEVEL/COURSE]. The goal is to generate a realistic, multi-perspective scenario that immerses students in a simulated real-world situation, allowing them to practice soft skills, navigate complex social dynamics, and deepen their conceptual understanding through embodied experience.

The guide must be structured as follows:

1. **The "Why" it Matters**: A brief, impactful explanation of the core principle behind role-play (e.g., "Role-play is a form of low-stakes simulation that builds empathy and procedural knowledge. It allows students to 'try on' different perspectives, practice and fail safely in difficult conversations, and emotionally internalize abstract concepts, leading to deeper and more durable learning than passive instruction.").

2. **Overarching Learning Objectives**: State the 2-3 primary knowledge, skill, and dispositional goals for this activity (e.g., "Knowledge: Understand the stakeholders and tensions in [TOPIC]. Skill: Practice active listening and persuasive communication. Disposition: Develop empathy for viewpoints different from one's own.").

3. **The Scenario Setup**: Provide all the context needed for students to step into their roles.

   ◇ **Scenario Title**: A engaging name for the exercise.

   ◇ **The Situation**: A paragraph setting the scene, explaining the context, and stating the central problem or dilemma to be resolved.

   ◇ **The Central Challenge**: The specific, urgent question or goal the group must address (e.g., "Your task is to decide whether to approve the new policy proposal by the end of the session.").

4. **The Roles**: Create 3-5 distinct roles. For each role, include:

   ◇ **Role Name & Title**: (e.g., "Dr. Ava Chen, Environmental Scientist")

◇ **Character Profile**: 2-3 sentences on their background, motivation, and primary goal in this scenario.

◇ **Secret Directive**: A private instruction or piece of information only that character knows. This is crucial for creating authentic tension and dialogue (e.g., "Secret: You have been promised funding for your lab if this proposal passes. You cannot reveal this to the others.")

◇ **Suggested Talking Points**: 2-3 ideas for what the character might say to advocate for their position.

5 **The Script & Flow**: Instead of a rigid line-by-line script, provide a flexible framework to guide the action.

◇ **Phase 1**: Individual Preparation (5 mins): Students read their role sheets and get into character.

◇ **Phase 2**: Open Discussion (10-15 mins): Characters meet and discuss the central challenge. The teacher acts as a facilitator, not a director.

◇ **Phase 3**: Negotiation & Resolution (5 mins): The group must attempt to reach a consensus, vote, or make a decision.

◇ **Potential Pathways**: Suggest 2-3 ways the conversation might logically go based on character motivations.

6 **The Debrief**: Guided Discussion (The Most Important Part): A list of 5-7 questions designed to process the experience metacognitively, moving from emotional response to conceptual understanding.

◇ **Level 1**: Reaction: (e.g., "How did it feel to be in your role? What was the most frustrating part?")

◇ **Level 2**: Analysis: (e.g., "What were the main conflicts? Why was it difficult to reach an agreement?")

◇ **Level 3**: Application: (e.g., "What does this teach us about real-world [TOPIC]? How will this change how you view news about [TOPIC]?")

7 **The "Why" Behind This Strategy**: A concluding summary explaining how this immersive, reflective process transforms students from passive learners into active participants, creating emotional anchors for memory and building essential life skills in conflict resolution and perspective-taking.

8 **Bonus**: Pro-Tip for Facilitation - The Teacher's Role:

◇ **Before**: "Brief participants on confidentiality and respect. This is a 'vulnerable' learning activity."

◇ **During**: "Your job is to observe and gently keep the conversation on track. If energy lags, inject a new piece of 'information' (e.g., 'The news has just reported that...')."

◇ **After**: "During debrief, protect students by discussing the characters and choices, not the students themselves. Ask, 'What did the Mayor do well?' rather than 'What did Sarah do well?'"

The final output should be a complete, plug-and-play simulation kit that provides a safe, structured, and powerful experiential learning opportunity, empowering students to learn through doing and reflecting.

Complete Output

Draft: Sample Role-Play Scenario Kit

Grade Level: 10 | Subject: Social Studies | Topic: Environmental Policy Debate

1. The "Why" it Matters

Role-play is a safe simulation that helps students explore conflict, empathy, and complex problem-solving. By taking on diverse roles in a realistic scenario, students learn to balance personal interest with community good — a core democratic skill.

2. Overarching Learning Objectives

- **Knowledge**: Understand the perspectives of stakeholders in environmental decision-making.
- **Skill**: Practice persuasive speaking, active listening, and negotiation.
- **Disposition**: Build empathy for different priorities in civic life.

3. The Scenario Setup

- **Scenario Title**: "The Future of Green Valley"
- **The Situation**: Green Valley is home to a rich ecosystem but also faces economic struggles. A company has proposed building a new factory that promises jobs but threatens local wetlands.
- **The Central Challenge**: By the end of the meeting, the town council (the student group) must vote whether to approve or reject the factory proposal.

4. The Roles

1. Dr. Maya Lopez – Environmental Scientist

- **Profile**: Has studied the wetland ecosystem for 15 years.
- **Secret**: Recently received grant funding that depends on continued biodiversity.
- **Talking Points**: "The wetlands purify our water," "Once destroyed, this ecosystem cannot recover."

2. Mr. Alan Brooks – Local Business Owner

- **Profile**: Owns a struggling store in town; sees the factory as a lifeline.
- **Secret**: His brother is applying for a job at the factory.
- **Talking Points**: "We need jobs now," "Without this investment, families will keep leaving town."

3. Samantha Kim – High School Student Representative

- **Profile**: Represents student council; cares about climate change but also sees peers struggling.
- **Secret**: Wants to run for class president and worries about losing support if she picks a side.
- **Talking Points**: "Young people deserve a voice," "This decision impacts our future most."

4. Mayor Jordan Smith – Town Mayor

- **Profile**: Responsible for the final decision; must balance economy, environment, and public opinion.
- **Secret**: The company has hinted at supporting his re-election campaign.
- **Talking Points**: "We must balance progress and preservation," "The town can't afford division."

5. The Script & Flow

- **Phase 1 (5 min)**: Individual role prep. Students read profiles, highlight goals, and get into character.

- **Phase 2 (10–15 min)**: Open council meeting. Each role presents opening statement, followed by open discussion.

- **Phase 3 (5 min)**: Negotiation & resolution. Group must vote or draft a compromise plan.

- **Potential Pathways**:

  ◇ A split vote leads to heated debate → mayor decides.

  ◇ Group proposes compromise (e.g., smaller factory, environmental safeguards).

  ◇ Factory rejected, leading to debate on alternative economic solutions.

6. The Debrief

- **Reaction**:

  ◇ "How did it feel to be in your role?"

  ◇ "What was frustrating about defending your character's position?"

- **Analysis**:

  ◇ "Which values clashed most strongly in this debate?"

  ◇ "Why was compromise difficult?"

- **Application**:

  ◇ "How does this mirror real-world debates on environment vs. economy?"

◇ "How might this change how you view news about climate policies?"

7. The "Why" Behind This Strategy

This scenario makes abstract tensions tangible. By embodying different stakeholders, students experience both the emotional stakes and the systemic complexity of decision-making, fostering deeper civic awareness and empathy.

8. Bonus: Pro-Tips for Facilitation

- **Before**: Remind students they are playing roles — disagreements are part of the scenario, not personal.
- **During**: If discussion slows, add a "breaking news update" (e.g., "New report: rare bird found in wetland").
- **After**: Frame reflections around roles, not individuals, to keep the focus on learning.

This kit is classroom-ready. Teachers can hand out role cards, set the timer, and run the activity with confidence.

Prompt: Icebreaker Activity Generator

Simple Prompt: "Act as an expert in 'Social-Emotional Learning (SEL)' and 'Educational Pedagogy'. Your task is to create a teacher-ready guide of purposeful icebreaker activities and questions that foster community and engagement".

- **Tone**: Fun, Friendly
- **Format**: Icebreaker Ideas (x3)
- **Platform**: First-Day Classes, Group Activities
- **AI Role**: Classroom Community Builder / Student Engagement Specialist
- **Prompt Goal**: Build trust and break down barriers among students
- **Tags**: #icebreaker, #studentEngagement, #classroomActivities
- **Prompt Variables**: {grade/level}, {classSize}, {timeAvailable}

Best Practice Tip: Choose activities that align with cultural sensitivity and age appropriateness.

Enhanced Prompt:

Act as an expert in 'Social-Emotional Learning (SEL)' and 'Educational Pedagogy'. Your task is to create a comprehensive list of icebreaker questions for a [GRADE LEVEL/COURSE] classroom. The goal is to generate questions that are strategically designed to do more than just break the silence; they must build genuine connections, establish psychological safety, activate prior knowledge, and set the tone for a collaborative and intellectually curious learning environment.

The guide must be structured as follows:

1 **The "Why" it Matters**: A brief, impactful explanation of the core principle behind intentional icebreakers (e.g., "The right icebreaker is a foundational investment in your classroom community. It is a high-leverage strategy to quickly build trust, normalize vulnerability, and signal to students that this is a space where their voice is heard and their experiences are valued. This initial investment pays dividends in increased participation, collaboration, and risk-taking throughout the entire year.").

2 **Overarching Goal**: State the primary objective of this icebreaker session (e.g., "To move beyond superficial introductions and foster authentic connections, activate students' cognitive and social-emotional readiness for learning, and collaboratively establish norms of respect and active listening.").

3 **The Icebreaker Menu**: A Strategic Mix Provide a list of 5-7 distinct questions. Ensure there is a deliberate mix of the following types to serve different purposes:

   ◇ **Community & Connection Builders**: To find common ground and build empathy.

   ◇ **Metacognitive & Reflective**: To get students thinking about their own thinking and learning.

   ◇ **Content-Based & Skill-Revealing**: To subtly introduce themes of the course and assess prior knowledge.

   ◇ **Creative & Imaginative**: To spark joy, lower inhibitions, and engage different parts of the brain.

4 **Breakdown for Each Icebreaker Question**: For every question, include:

◇ **The Question**: Phrased precisely as the teacher should ask it.

◇ **Icebreaker Type**: Categorize it from the list above (e.g., "Creative & Imaginative").

◇ **The "Why" it's Important**: A brief explanation of the specific social-emotional or academic skill it targets and its long-term impact (e.g., "This question requires students to make connections and synthesize information, a key critical thinking skill. It also reveals diverse perspectives and normalizes that there is no single 'right' answer in this classroom.").

◇ **Implementation Tip**: Clear, practical advice for the teacher on how to facilitate it (e.g., "Use a Turn-and-Talk protocol first to ensure all students have time to think," "Model your answer first to provide a scaffold and build vulnerability.").

5 **The "Why" Behind This Strategy**: A concluding summary explaining how this intentional approach to icebreakers accelerates the formation of a cohesive learning community, provides the teacher with invaluable insights into student personalities and readiness, and creates a classroom culture where students feel safe to engage deeply with content and with each other.

6 **Bonus**: Pro-Tip for Teachers - Facilitation for Safety & Inclusion:

◇ **The Right to Pass**: Always state that students have the "right to pass" if they are not comfortable answering a question. This builds trust.

◇ **Teacher Participation**: Always answer the question yourself first. This models vulnerability, length of response, and tone.

◇ **Think-Pair-Share**: For deeper questions, use this structure. This gives everyone a low-risk chance to speak before sharing with the whole group.

◇ **Debrief the Process**: Afterward, ask, "What did we learn about our class from this activity?" This metacognitive step solidifies the community building.

The final output should be a complete, plug-and-play icebreaker kit that provides a clear path from awkward silence to a connected, engaged classroom ready to learn.

Complete Output:

The "Why" it Matters: The first few minutes of a class set the emotional and intellectual tone for the entire year. A well-chosen icebreaker is a diagnostic tool, a community-building exercise, and a values-setting activity all in one. It tells students they are seen as whole people, not just repositories for information, which is the first step toward creating a truly inclusive learning environment.

The Icebreaker Menu: Questions with Purpose

1. Question: "If you were to create a museum exhibit about one thing you're passionate about, what would it be and what would be the first item in the exhibit?"

- **Type**: Creative & Imaginative / Community Builder
- **Why It's Important**: This question moves beyond "what's your hobby?" to uncover deeper passions and values. It allows students to showcase expertise in a non-academic area, building confidence and allowing

peers to see them in a new light. It also taps into creativity and descriptive language.

- **Implementation Tip**: Model this with your own passion (e.g., "Mine would be on vintage typewriters, and the first item would be my great-grandfather's 1920 Underwood"). Use Think-Pair-Share to let everyone brainstorm before sharing with the whole group.

2. Question: "What is a 'superpower' you have that helps you learn? (It could be anything: asking great questions, making cool diagrams, being a great listener, etc.)"

- **Type**: Metacognitive & Reflective
- **Why It's Important**: This frames learning strengths as superpowers, which is empowering and positive. It forces metacognitive reflection from day one and begins to create a shared vocabulary around learning strategies. It also allows you to quickly identify potential collaborators and understand the diverse strengths in your room.
- **Implementation Tip**: Chart these on the board as students share. This visually validates their strengths and shows the collective power of the classroom.

3. Question: "What is one question you have about [COURSE SUBJECT] that you've always wanted to answer?" (e.g., "about space," "about history," "about how stories are written")

- **Type**: Content-Based & Skill-Revealing
- **Why It's Important**: This instantly frames your subject as a field of active inquiry, not a static set of facts. It activates prior knowledge and curiosity, and it provides you with a goldmine of information about student interests, misconceptions, and readiness that you can use to tailor your units.

- **Implementation Tip**: Record these questions on an anchor chart titled "Our Investigating Board" and return to them throughout the year as they are answered.

4. Question: "Complete this sentence: 'A perfect learning environment is one where…'"

- **Type**: Community Builder / Metacognitive
- **Why It's Important**: This is a covert way to collaboratively set classroom norms and expectations. Students are literally defining the culture they want to be a part of. It gives them agency and ownership over the classroom environment from the very beginning.
- **Implementation Tip**: Chart the responses and use them to co-create your class's list of norms or agreements. For example, if a student says "…where it's okay to make mistakes," you can propose that as a core norm.

5. Question: "What is one small thing that always makes your day better?"

- **Type**: Community & Connection Builder
- **Why It's Important**: This question is simple, safe, and universally accessible. It focuses on small joys, fostering positivity and empathy. It often reveals commonalities (e.g., "a sunny day," "my dog," "a good snack") that quickly build a sense of shared experience and community.
- **Implementation Tip**: This is a great quick warm-up or closing question. It's low-stakes and ends things on a positive note.

The **"Why" Behind This Strategy**: This plan works because it treats community building as academic work. These questions are designed to elicit responses that are inherently interesting and revealing, allowing students to connect on a human level while simultaneously practicing the kinds of thinking—metacognition, inquiry, creativity—that you will value all year long. This seamless integration ensures the activity feels purposeful, not perfunctory.

Bonus: Pro-Tip for Teachers - Facilitation for Safety & Inclusion:

- **Never Force Whole-Class Sharing**: Use small groups or pairs (Turn-and-Talk) for the initial response. Then, you can ask, "Who heard something fascinating from their partner they'd like to share?" This takes the pressure off and encourages active listening.

- **Be Mindful of Trauma**: Avoid questions that could inadvertently trigger painful memories (e.g., "What was the best vacation you ever took?" might be difficult for some students). Stick to questions focused on the present, preferences, and learning.

- **Use a Talking Stick**: For whole group sharing, use an object to signify whose turn it is to speak. This manages conversation and reinforces respectful listening.

Prompt: Classroom Scavenger Hunt Ideas

Simple Prompt: "Act as a 'Classroom Gamification Coach' and 'Experiential Learning Specialist'. Your task is to create a teacher-ready list of curriculum-aligned scavenger hunt activities for a [GRADE LEVEL] classroom.".

- **Tone**: Playful, Interactive
- **Format**: Scavenger Hunt Plan
- **Platform**: Classroom, School Events
- **AI Role**: Classroom Gamification Coach / Experiential Learning Specialist
- **Prompt Goal**: Combine fun and learning through interactive exploration
- **Tags**: #scavengerHunt, #gamification, #studentEngagement
- **Prompt Variables**: {topic/subject}, {grade/level}, {learningGoals}

Best Practice Tip: Use both physical and digital clues (like QR codes) to increase engagement.

Enhance Prompt:

Act as a 'Classroom Gamification Coach' and 'Experiential Learning Specialist'. Your task is to create a comprehensive list of scavenger hunt ideas for a [GRADE LEVEL] classroom. The goal is to generate ideas that transform a simple game into a powerful pedagogical tool, promoting collaboration, critical thinking, problem-solving, and the active application of knowledge in a dynamic and memorable way.

The guide must be structured as follows:

1 **The "Why" it Matters**: A brief, impactful explanation of the core principle behind educational scavenger hunts (e.g., "A well-designed scavenger hunt is the ultimate fusion of engagement and education. It leverages the innate human drive for play, discovery, and achievement to motivate students to actively retrieve information, apply skills, and collaborate with peers. This transforms passive learning into an immersive experience, dramatically increasing retention and making abstract concepts tangible.").

2 **Overarching Goal**: State the primary objective of this activity (e.g., "To reinforce key unit concepts, familiarize students with their learning environment, foster team-building, and assess understanding in a low-stakes, high-energy format that celebrates curiosity and problem-solving.").

3 **The Scavenger Hunt Menu**: A Spectrum of Learning Provide a list of 5-7 distinct hunt ideas. Ensure there is a deliberate mix of the following types to serve different pedagogical purposes:

   ◇ **Getting-to-Know-You**: To build community and familiarize students with the classroom.

   ◇ **Content Review & Mastery**: To reinforce and apply academic knowledge.

   ◇ **Skill-Based Application**: To practice specific skills like reading, math, or digital literacy.

   ◇ **Just-for-Fun & Team Building**: To boost morale and reinforce classroom culture.

4 **Breakdown for Each Scavenger Hunt Idea**: For every idea, include:

◇ **Hunt Title & Type**: A catchy, descriptive name and category (e.g., "Textbook Explorer (Content Review)").

◇ **Core Concept**: A 1-2 sentence description of the theme and goal.

◇ **The "Why" it's Important**: A brief explanation of the specific learning or social-emotional skill it targets (e.g., "This hunt moves vocabulary study from a solitary task to a collaborative discovery process, creating stronger neural connections through physical movement and social interaction.").

◇ **Sample Clue/Task**: A specific example of what students would be asked to find or do.

◇ **Practical Tip**: Advice for seamless execution (e.g., "Use QR codes to reveal clues," "Have teams take photos of their finds as proof.").

5 **The "Why" Behind This Strategy**: A concluding summary explaining how this gamified, active learning approach is more effective for memory and engagement than traditional review methods. It caters to kinesthetic and social learners, provides immediate feedback, and allows the teacher to act as a facilitator observing student thinking in action.

6 **Bonus**: Pro-Tip for Teachers - Facilitation & Flow:

◇ **Clear Rules First**: Always review the rules, boundaries, and noise level expectations before starting.

◇ **Strategic Grouping**: Pre-make groups to mix abilities and personalities. Designate roles (e.g., Reader, Recorder, Runner, Photographer) to ensure full participation.

⋄ **The Debrief is Key**: Allocate time at the end for teams to share their findings and strategies. This solidifies the learning and allows you to clarify any misconceptions.

⋄ **Have a "Puzzle Master" Box**: For teams that finish early, have a challenge box with a bonus riddle or problem related to the hunt to solve.

The final output should be a complete, plug-and-play scavenger hunt kit that provides a clear path from concept to exciting execution.

Complete Output:

The "Why" it Matters: Scavenger hunts capitalize on the power of active retrieval practice—one of the most effective learning strategies known. By forcing students to recall information and apply it in a new context to solve a problem, they build much stronger and more durable memory pathways than through passive review. The game format lowers the affective filter, encouraging risk-taking and making learning feel like an adventure.

The Scavenger Hunt Menu: Adventures in Learning

1. Hunt: "Classroom Basics" Bootcamp

- **Type**: Getting-to-Know-You / Procedural
- **Core Concept**: A hunt designed to familiarize students with the physical layout and procedures of the classroom during the first week of school.
- **Why It's Important**: It teaches routines and locations (e.g., where to find extra paper, turn in work, get supplies) in an engaging way, promoting independence and reducing future questions.

- **Sample Clue/Task**: "Find the tray where you would turn in a completed essay. Write down the color of that tray."
- **Practical Tip**: Make the final clue lead to a small prize or a certificate proclaiming them "Official Classroom Experts."

2. Hunt: "Textbook Explorer"

- **Type**: Content Review & Mastery / Skill-Based
- **Core Concept**: Teams use their textbook or class readings to find answers to questions or locate specific features (e.g., index, glossary, specific chart).
- **Why It's Important**: It teaches crucial text navigation skills and reinforces unit-specific content simultaneously. It shows students that the textbook is a resource for finding answers.
- **Sample Clue/Task**: "On what page does the chapter about the water cycle begin? Find and write down the definition of 'condensation' from the glossary."
- **Practical Tip**: Use this before a test as a dynamic review session. It's far more engaging than a study guide.

3. Hunt: "QR Code Question Quest"

- **Type**: Content Review & Mastery / Tech Integration
- **Core Concept**: QR codes are posted around the room or school. Each code, when scanned, reveals a question or problem teams must solve.
- **Why It's Important**: It integrates technology seamlessly, feels modern and exciting for students, and allows you to easily differentiate by assigning different leveled questions to different codes.
- **Sample Clue/Task**: (The QR code links to a Google Form question): "Solve for x: $2x + 5 = 17$. Submit your

answer here."

- **Practical Tip**: Use a free QR code generator. The Google Form can collect all team answers in a spreadsheet for easy grading.

4. Hunt: "Photo Proof" Challenge

- **Type**: Skill-Based Application / Creative
- **Core Concept**: Instead of writing answers, teams must take photographs that meet specific criteria.
- **Why It's Important**: It fosters creativity, perspective-taking, and visual literacy. It's excellent for assessing understanding of abstract concepts like shapes, angles, or verbs.
- **Sample Clue/Task**: "Find and photograph an example of a right angle somewhere in this classroom." or "Stage a photo that demonstrates the meaning of the word 'collaboration'."
- **Practical Tip**: Create a shared Google Photos album where each team can upload their pictures for a final gallery walk.

5. Hunt: "Primary Source Detective"

- **Type**: Content Review & Mastery (Social Studies Focus)
- **Core Concept**: For a history unit, teams analyze different primary sources (e.g., maps, letters, political cartoons, artifacts) placed around the room to answer questions.
- **Why It's Important**: It builds critical historical thinking skills—sourcing, contextualization, and corroboration—in a hands-on, inquiry-based way.
- **Sample Clue/Task**: "At Station 3, analyze the political cartoon. Who do the two characters represent? What is the artist's opinion on this issue?"

- **Practical Tip**: Use clear plastic sleeves to protect delicate documents or images.

The "Why" Behind This Strategy: This plan works because it disguises rigorous learning as play. The hunt format inherently differentiates—stronger students can lead the problem-solving, while others contribute in different ways. It provides the teacher with real-time data on which concepts teams struggle with (by where they get stuck) and creates a vibrant, energetic learning atmosphere where students are fully invested in the process of discovery.

Bonus: Pro-Tip for Teachers - Facilitation & Flow:

- **The Answer Key Station**: Have a central "check-in" station where a team must bring their answer sheet to you for verification before getting the next clue. This prevents a runaway team and lets you catch errors early.

- **Set a Timer**: A visible timer adds excitement and helps manage the energy and pace of the activity.

- **Embrace the Productive Noise**: A scavenger hunt is supposed to be active and a bit noisy. This is a sign of engagement, not misbehavior.

Prompt: Classroom Bulletin Board Ideas

Simple Prompt: "Act as a 'Learning Environment Designer' and 'Classroom Visual Learning Specialist'. Your task is to create a teacher-ready menu of purposeful bulletin board ideas that enhance learning, community, and student ownership for a [Subject/GRADE LEVEL] classroom".

- **Tone**: Creative, Inspiring
- **Format**: Idea List (x5)
- **Platform**: Classroom Environment
- **AI Role**: Learning Environment Designer / Classroom Visual Learning Specialist
- **Prompt Goal**: Enhance learning spaces with engaging visual aids
- **Tags**: #bulletinBoards, #visualLearning, #classroomDesign
- **Prompt Variables**: {subject}, {grade/level}, {theme}

Best Practice Tip: Allow students to contribute content (quotes, drawings) to make boards more participatory.

Enhanced Prompt:

Act as a 'Learning Environment Designer' and 'Classroom Visual Learning Specialist'. Your task is to create a comprehensive menu of classroom bulletin board ideas for a [GRADE LEVEL] classroom. The goal is to generate ideas that transform bulletin boards from static decorations into dynamic, interactive learning tools that reinforce curriculum, celebrate growth, build community, and actively engage students in their own learning process.

The guide must be structured as follows:

1 **The "Why" it Matters**: A brief, impactful explanation of the core principle behind using bulletin boards as a strategic teaching tool (e.g., "A well-designed bulletin board is a high-impact learning station, not just wall decoration. It acts as a 'silent teacher'—providing constant visual reinforcement of key concepts, anchoring instruction, documenting the learning journey, and inviting student interaction. It makes thinking visible and celebrates the process of learning, not just the final product.").

2 **Overarching Goal**: State the primary objective of this approach (e.g., "To create a print-rich environment that supports independent learning, fosters a sense of belonging and pride, and serves as a constant, engaging reference point for core skills and knowledge throughout a unit or the entire year.").

3 **The Bulletin Board Menu**: A Strategic Mix Provide a list of 5-7 distinct board ideas. Ensure there is a deliberate mix of the following types to serve different pedagogical purposes:

   ◇ **Interactive & Participatory**: Boards students can touch, move, and contribute to.

   ◇ **Anchor Charts & Reference**: Boards that store essential, frequently accessed knowledge.

   ◇ **Celebratory & Documentary**: Boards that showcase student work and process.

   ◇ **Community & Culture Building**: Boards that foster belonging and relationships.

4 **Breakdown for Each Bulletin Board Idea**: For every idea, include:

◇ **Board Title**: A catchy, descriptive name.

◇ **Core Concept**: A 1-2 sentence description of the board's purpose and setup.

◇ **The "Why" it's Important**: A brief explanation of the specific learning or social-emotional skill it targets and its long-term impact (e.g., "This board moves vocabulary from a static list to a living, growing part of the classroom, encouraging word acquisition through playful interaction and visual reinforcement.").

◇ **Key Components**: What elements are needed to create it (e.g., "Student work samples," "QR codes," "Question prompts," "Pockets with cards," "Laminated pieces").

◇ **Practical Tip**: One piece of advice for easy execution and durability (e.g., "Use a sturdy fabric background instead of paper—it doesn't fade or tear and can last all year," "Laminate pieces so students can write on them with dry-erase markers.").

5 **The "Why" Behind This Strategy**: A concluding summary explaining how this intentional approach to bulletin boards—prioritizing interactivity, reference, and celebration—creates a more effective and stimulating learning environment than traditional, teacher-made decorative boards. It empowers students, supports independence, and makes learning tangible.

6 **Bonus**: Pro-Tip for Teachers - The Workflow:

◇ **Student-Created is Best**: The most powerful boards are co-created with students. This builds ownership and saves you time.

◇ **Less is More**: Avoid visual clutter. A board with a clear focus and plenty of blank space is more effective than one crammed with information.

◇ **Use a Digital Tool**: Use Canva or Google Slides to design and print headings, labels, and graphics quickly and professionally.

◇ **Designate a Curator**: Assign a student job like "Board Manager" to help maintain and update interactive boards.

The final output should be a complete, plug-and-play bulletin board kit that provides a clear vision and practical steps for creating learning-focused classroom displays.

Complete Output:

The "Why" it Matters: The classroom environment is a primary teaching resource. A strategic bulletin board reduces the need for constant re-teaching by serving as a permanent visual anchor. It answers the questions "What are we learning?" and "Why does it matter?" and empowers students to find answers themselves, fostering resourcefulness and independence.

The Bulletin Board Menu: From Decoration to Education

1. Idea: Vocabulary Vine (or Word Wall)

- **Type**: Interactive & Reference
- **Core Concept**: A growing display of key unit vocabulary. Each word is on a separate leaf, card, or strip, with a student-friendly definition and a visual cue.
- **Why It's Important**: It provides constant visual exposure to essential terminology, which is crucial for literacy across all subjects. The "growing" nature shows learning progression.

- **Key Components**: Sturdy background, die-cut leaves or cards, markers, pins or velcro dots.
- **Practical Tip**: Have students add new words as they are learned. For an interactive twist, include QR codes that link to videos or definitions.

2. Idea: "We Are Experts" Spotlight Board

- **Type**: Celebratory & Documentary
- **Core Concept**: A dedicated space to showcase exemplary student work, not just final products but also drafts, sketches, and reflections that show the learning process.
- **Why It's Important**: It validates student effort, provides authentic models of high-quality work, and builds a culture of pride and shared accomplishment. It tells students, "Your work is important and worth displaying."
- **Key Components**: Black paper or a frame to border work, clothespins for easy rotation, a small placard for the student to explain their work.
- **Practical Tip**: Rotate work weekly. Ensure every student gets featured over time. Use this board to introduce success criteria for assignments.

3. Idea: "What Are We Investigating?" Inquiry Board

- **Type**: Interactive & Participatory
- **Core Concept**: The central hub for a current unit. It features the unit's essential question, anchor charts, student questions on sticky notes, and real-world photos or artifacts.
- **Why It's Important**: It builds anticipation and frames the entire unit around inquiry. The sticky notes allow all students to share their wonders and see their curiosity driving the class's learning.

- **Key Components**: Large central question, anchor charts, pockets of sticky notes, relevant images or realia.
- **Practical Tip**: Start the unit with the board mostly blank. Add the anchor charts with the students during lessons, making the board a living record of your collective learning.

4. Idea: "Help Yourself" Resource Board

- **Type**: Reference
- **Core Concept**: A board dedicated to self-service learning tools. It includes things like a writing checklist, a multiplication table, a map, a timeline, a "transition words" poster, and emotion regulation strategies.
- **Why It's Important**: This is the ultimate tool for fostering executive functioning and independence. It teaches students to resource themselves before asking for help, freeing up the teacher to support students with more complex needs.
- **Key Components**: Laminated reference sheets on rings or in pockets, clearly labeled sections.
- **Practical Tip**: Explicitly teach students how and when to use every resource on this board at the beginning of the year.

5. Idea: "Kindness Catch" Community Board

- **Type**: Community & Culture Building
- **Core Concept**: A board where students can publicly acknowledge acts of kindness they witness. They write a note ("I caught [Name] being kind when they...") and pin it up.
- **Why It's Important**: It proactively reinforces positive peer interactions and builds a culture of empathy, gratitude, and noticed kindness. It shifts the focus from catching misbehavior to catching kindness.

- **Key Components**:"I Caught You Being Kind" slips, a designated basket for submissions, plenty of push pins.

- **Practical Tip**: Read a few "catches" aloud during morning meeting or closing circle to amplify their positive impact.

The "Why" Behind This Strategy: This plan works because it treats wall space as valuable instructional real estate. By choosing boards that are interactive, reference-based, or celebratory, you ensure they are constantly working for you and your students. This approach supports differentiated learning, reduces repetitive questions, and creates a classroom environment that is both intellectually stimulating and emotionally supportive.

Bonus: Pro-Tip for Teachers - The Workflow:

- **Batch Your Boards**: Design and create all your background boards (fabric, borders, titles) before the school year starts. Then you can just switch out the content.

- **Use Rubric Charts**: Create anchor charts that are essentially rubrics for major assignments (e.g., "What does a 4-star paragraph include?") and keep them on the board for the duration of the project.

- **Go 3D**: Use the board's ledge for related objects, student-built models, or a "question box" to make it more dynamic.

- **It's Okay to Have a "Pretty" Board**: One board can be purely seasonal or inspirational. A visually calming and pleasant environment also supports learning. The key is intentionality.

Prompt: Educational App Generator

Simple Prompt: "Act as an 'EdTech Strategist' and 'Digital Learning Designer'. Your task is to create a teacher-ready guide of educational applications that directly support curriculum goals and student engagement for [SUBJECT/TOPIC] designed for [GRADE LEVEL]".

- **Tone**: Innovative, Practical
- **Format**: App Concept Ideas (x3)
- **Platform**: Edtech, Classroom Tools
- **AI Role**: EdTech Strategist / Digital Learning Designer
- **Prompt Goal**: Inspire digital tools that support modern teaching
- **Tags**: #edtech, #apps, #studentLearning
- **Prompt Variables**: {topic}, {grade/level}, {learning-Objectives}

Best Practice Tip: Consider accessibility features so apps can be inclusive for all students.

Enhanced Prompt:

Act as an 'EdTech Strategist' and 'Digital Learning Designer'. Your task is to create a comprehensive guide to educational applications for [SUBJECT/TOPIC] designed for [GRADE LEVEL]. The goal is to curate a list of apps that are not just "fun" but are strategically chosen to enhance, extend, and transform student learning by providing personalized practice, fostering creativity, and offering authentic assessment data.

The guide must be structured as follows:

1 **The "Why" it Matters**: A brief, impactful explanation of the core principle behind effective EdTech integration (e.g., "Technology is not a substitute for teaching; it is a powerful multiplier. The right app, used at the right time, can provide individualized, immediate feedback, unlock new forms of creative expression, and make abstract concepts visually tangible and interactive, thereby deepening understanding and personalizing the learning journey.").

2 **Overarching Learning Objectives**: State the 2-3 primary instructional goals this technology is designed to support (e.g., "To provide differentiated skill practice," "To facilitate collaborative project-based learning," "To enable students to create and share multimodal demonstrations of understanding.").

3 **The App Menu**: A Strategic Toolkit Provide a list of 5-7 distinct applications. Ensure there is a deliberate mix of the following types to serve different pedagogical purposes:

   ◇ **Skill & Drill**: For adaptive practice and fluency building.

   ◇ **Content Creation & Curation**: For students to demonstrate understanding through multimedia.

   ◇ **Simulation & Exploration**: For interacting with complex systems or concepts.

   ◇ **Collaboration & Communication**: For fostering teamwork and discussion.

   ◇ **Formative Assessment & Feedback**: For providing real-time data on student understanding.

4 **Breakdown for Each Application**: For every app,

include:

- ◇ **App Name & Platform**: (e.g., "Kahoot! (Web, iOS, Android)").

- ◇ **Category**: Label it from the list above (e.g., "Formative Assessment & Feedback").

- ◇ **Core Function**: A 1-2 sentence description of what it does.

- ◇ **The "Why" it's Important**: A brief explanation of the specific learning benefit and long-term impact (e.g., "This tool transforms review sessions into high-energy, low-stakes competitions, dramatically increasing engagement and providing the teacher with instant, whole-class data on which concepts need re-teaching.").

- ◇ **Ideal Use Case**: A specific example of when and how to integrate it into a lesson (e.g., "Perfect for a unit review session or as a quick bell-ringer to activate prior knowledge.").

5 **The "Why" Behind This Strategy**: A concluding summary explaining how this curated, purpose-driven approach to technology prevents "app fatigue," ensures screen time is meaningful and productive, and empowers teachers to make informed decisions that directly support their instructional goals.

6 **Bonus**: Pro-Tip for Teachers - The Integration Framework:

- ◇ **The SAMR Model**: Use this as a filter: Does the app simply Substitute a worksheet (okay), or does it Augment, Modify, or Redefine the learning task (ideal)?

◇ **Data-Driven Decisions**: Choose apps that provide you with actionable data on student progress, not just a final score.

◇ **Balance is Key**: Technology is one tool in the toolbox. The best learning environments blend tech-enabled activities with hands-on, un-plugged, and social learning.

◇ **Student Choice**: Where possible, offer a "menu" of apps for students to choose from to demonstrate their understanding, fostering own-ership and creativity.

The final output should be a complete, plug-and-play EdTech integration guide that is easy to personalize and provides clear, curriculum-aligned value.

Complete Output:

The "Why" it Matters: In a world of digital distraction, the educational value of an app is determined by its pedagogical purpose. The best apps are those that act as force multipliers—they allow us to do things that were previously impossible in the classroom, such as offer truly personalized learning paths, facilitate global collaboration, or provide immersive experiences that bring complex ideas to life. This shifts technology from being a babysitter to being a bridge to deeper understanding.

The App Menu: A Strategic Toolkit for [SUBJECT]

1. App: Khan Academy

- **Category**: Skill & Drill / Tutorial
- **Core Function**: Provides a vast library of practice exercises, instructional videos, and a personalized learning dashboard.

- **Why It's Important**: It offers mastery-based learning, allowing students to work at their own pace to fill gaps in understanding or accelerate ahead. It provides teachers with unparalleled data on individual and class-wide progress.

- **Ideal Use Case**: Station rotation for differentiated practice, flipped classroom model, or targeted intervention for struggling students.

2. App: Quizlet

- **Category**: Skill & Drill

- **Core Function**: Allows users to create and study digital flashcards, with built-in games and practice tests.

- **Why It's Important**: It leverages spaced repetition and active recall, two of the most effective study techniques proven by cognitive science. The game modes (like Match and Gravity) make vocabulary and fact memorization engaging and efficient.

- **Ideal Use Case**: Vocabulary building, reviewing key terms and facts, student-created study sets for unit tests.

3. App: Flip (formerly Flipgrid)

- **Category**: Collaboration & Communication / Formative Assessment

- **Core Function**: A video discussion platform where teachers post prompts and students respond with short videos.

- **Why It's Important**: It amplifies student voice, especially for shy or reluctant writers. It builds a community of learners and allows for authentic, verbal demonstration of understanding. It also helps develop communication and digital citizenship skills.

- **Ideal Use Case**: Exit tickets, book talks, peer feedback sessions, explaining problem-solving steps orally.

4. App: Canva for Education

- **Category**: Content Creation & Curation
- **Core Function**: A simple, powerful graphic design tool with thousands of templates for presentations, infographics, posters, and videos.
- **Why It's Important**: It allows students to demonstrate their understanding visually and creatively, moving beyond the traditional essay or report. This fosters digital literacy and design thinking skills highly relevant to the modern world.
- **Ideal Use Case**: Creating science fair posters, designing book covers for a novel study, producing public service announcements, making data visualizations.

5. App: Google Jamboard

- **Category**: Collaboration & Communication
- **Core Function**: A cloud-based, collaborative digital whiteboard.
- **Why It's Important**: It facilitates real-time, visual collaboration. Students can brainstorm, map ideas, and problem-solve together from any device, capturing the energy of a group whiteboard session in a digital, savable format.
- **Ideal Use Case**: Brainstorming sessions, Venn diagrams, mind mapping, group problem-solving in math, "gallery walks" to view each other's work.

6. App: PhET Interactive Simulations (University of Colorado Boulder)

- **Category**: Simulation & Exploration
- **Core Function**: Provides free, research-based science and math simulations.
- **Why It's Important**: It makes abstract, invisible, or

dangerous concepts (like gravity, circuit building, or natural selection) tangible, visual, and safe for students to manipulate and explore. This builds deep conceptual understanding through inquiry.

- **Ideal Use Case**: Introducing a new physics concept, conducting virtual labs where physical materials are unavailable, allowing students to form and test hypotheses.

The "Why" Behind This Strategy: This plan works because it treats technology as a strategic partner, not a novelty. By carefully matching the app's function to a specific learning goal, you ensure that technology use is always intentional and effective. This curated approach saves teachers from the overwhelm of the app store and provides a versatile toolkit to support differentiation, engagement, and authentic assessment throughout a unit.

Bonus: Pro-Tip for Teachers - The Integration Framework:

- **SAMR in Action**: Ask: Can this app Redefine the task? Instead of a poster (Substitution), could students use Canva to create an infographic that is published online for a real audience (Modification/Redefinition)?

- **Manage the Workflow**: Use a platform like Google Classroom or Seesaw as a "home base" to distribute links and collect student work from various apps. This simplifies the process for everyone.

- **Always Have a Non-Tech Backup**: Technology can fail. Always have an analogous unplugged activity ready to go to avoid losing instructional momentum.

Prompt: Classroom Decoration Ideas

Simple Prompt: "Act as a 'Learning Environment Designer' and 'Classroom Space Architect'. Your task is to create a purposeful guide of classroom environment design ideas that align with [THEME] and reinforce student learning, engagement, and community around the theme of [THEME] for [GRADE LEVEL]".

- **Tone**: Creative, Supportive
- **Format**: Idea List (x5)
- **Platform**: Classroom Setup
- **AI Role**: Learning Environment Designer / Classroom Space Architect
- **Prompt Goal**: Create an engaging environment that reinforces curriculum
- **Tags**: #classroomDesign, #visualLearning, #studentEngagement
- **Prompt Variables**: {subject/topic}, {grade/level}

Best Practice Tip: Rotate decorations monthly to keep the classroom fresh and stimulating.

Enhanced Prompt:

Act as a 'Learning Environment Designer' and 'Classroom Space Architect'. Your task is to create a comprehensive guide to designing a classroom environment around the theme of [THEME] for [GRADE LEVEL]. The goal is to generate ideas that transform the physical space into a "third teacher," where the walls, boards, and displays actively contribute to student learning, foster a sense of identity and belonging, and stimulate curiosity and engagement.

The guide must be structured as follows:

1 **The "Why" it Matters**: A brief, impactful explanation of the core principle behind intentional classroom design (e.g., "The classroom environment is a silent curriculum. It subconsciously communicates values, sets expectations, and directly influences student behavior and engagement. A purposefully designed space isn't just decorative; it is a dynamic, interactive tool that reduces anxiety, supports executive functioning, and makes learning visible and celebrated.").

2 **Overarching Design Goal**: State the primary objective for the themed environment (e.g., "To create a space that immerses students in the [THEME], supports literacy and numeracy through constant visual reference, and showcases student voice and ownership, making the classroom truly theirs.").

3 **The Design Menu**: A Strategic Mix of Elements Provide a list of 5-7 distinct decoration ideas. Ensure there is a deliberate mix of the following types to serve different purposes:

◇ **Anchor Charts & Reference Walls**: For storing essential, frequently accessed knowledge.

◇ **Interactive & Participatory Displays**: For allowing students to manipulate and contribute to the environment.

◇ **Student Work Galleries**: For building community, pride, and metacognition.

◇ **Atmosphere & Ambiance Elements**: For creating a specific mood and stimulating curiosity.

◇ **Functional & Organizational Systems**: For promoting independence and reducing cognitive load.

4 **Breakdown for Each Decoration Idea**: For every idea, include:

◇ **Idea Title & Type**: A descriptive name and category (e.g., "Vocabulary Vine (Interactive Display)").

◇ **Core Concept**: A 1-2 sentence description of the element and its placement.

◇ **The "Why" it's Important**: A brief explanation of the pedagogical purpose and long-term impact (e.g., "This moves vocabulary from a static list to a living, growing part of the classroom, encouraging word acquisition through playful interaction and visual reinforcement.").

◇ **Materials Needed**: A simple, affordable list of items to create it (e.g., "Butcher paper, markers, student-generated word cards, tape").

◇ **Making it Interactive**: A suggestion for how students can use or contribute to it.

5 **The "Why" Behind This Strategy**: A concluding summary explaining how this intentional, student-centered approach to design creates a more inclusive and effective learning environment than pre-packaged, teacher-made decor. It empowers students, reduces off-task behavior by providing clear resources, and turns the classroom itself into a catalyst for inquiry.

6 **Bonus**: Pro-Tip for Teachers - Design on a Dime:

◇ **Student-Created is Best**: The most powerful

decor is made by students. It boosts ownership and saves you countless hours of labor.

◇ **Less is More**: Avoid visual clutter. Aim for "calm and purposeful," not "sensory overload." Designated, blank space is crucial for preventing anxiety and allowing the important elements to stand out.

◇ **Use What You Have**: Repurpose old frames, use fabric instead of paper for bulletin boards (it doesn't fade or tear), and harness natural light and plants to warm up the space.

◇ **The Power of Light**: Use lamps and fairy lights to create soft, calming pools of light instead of harsh overhead fluorescents. This dramatically changes the room's mood.

The final output should be a complete, plug-and-play design kit that empowers a teacher to create a classroom that is not only visually cohesive but also functionally dedicated to deepening student learning and community.

Complete Output

THEME: Exploration & Discovery

The "Why" it Matters: A classroom's design is its first message to students. A theme of "Exploration & Discovery" tells them that this is a place for curiosity, for asking questions, and for venturing into the unknown. Every element should serve as a map, a tool, or a record of their intellectual journey, making the process of learning as visible and celebrated as the final product.

The Design Menu: Transforming Your Classroom into a Basecamp for Exploration

1. Idea: The "Question of the Week" Launch Pad

- **Type**: Interactive & Participatory Display

- **Core Concept**: A designated bulletin board or wall space that features a new, open-ended, theme-related question each week (e.g., "What would you pack for a journey to the bottom of the ocean?"). Students post their answers on sticky notes around the question.

- **Why It's Important**: This cultivates a culture of inquiry and values every student's voice. It provides a constant, low-stakes writing prompt and builds the habit of wondering and hypothesizing.

- **Materials Needed**: Bulletin board paper, markers, a stash of sticky notes in various colors.

- **Making it Interactive**: Students add their responses throughout the week. A student can be chosen to be the "Question Curator" who reads the answers and shares a few highlights.

2. Idea: Anchor Chart Harbor

- **Type**: Anchor Charts & Reference Walls

- **Core Concept**: A dedicated, easy-to-see wall space where your most important instructional anchor charts are "docked." Charts are created with students during lessons and remain up for constant reference.

- **Why It's Important**: This supports metacognition and independence. Instead of relying solely on the teacher, students learn to use tools around them to recall strategies, processes, and key information, building executive functioning skills.

- **Materials Needed**: Large chart paper, bold markers, magnets or tape.

- **Making it Interactive**: During lessons, ask students, "Which chart could help us with this problem?" Encourage them to get up and point to the specific part that helps them.

3. Idea: "Cartographer's Corner" - Student Work Gallery

- **Type**: Student Work Galleries
- **Core Concept**: A display area framed as a place to "map our learning." It features rotating exemplary student work (projects, writing, art) that demonstrates exploration, risk-taking, and deep thinking.
- **Why It's Important**: This validates student effort and provides authentic models of high-quality work. It builds a culture of pride and shared accomplishment, showing that everyone's "discoveries" are valued.
- **Materials Needed**: Simple black frames, clothespins and string for a rotating display, a "Curator" label for the student whose work is featured.
- **Making it Interactive**: Students write a brief "Discovery Statement" to accompany their displayed work, explaining their process or what they learned.

4. Idea: "Specimen Station" - Curiosity Table

- **Type**: Atmosphere & Ambiance Element
- **Core Concept**: A small table with a magnifying glass, microscope (if available), and rotating natural or theme-related objects (e.g., fossils, geodes, unique shells, historical artifacts replicas).
- **Why It's Important**: This stimulates natural curiosity and scientific observation skills. It provides a purposeful activity for early finishers and caters to kinesthetic and visual learners, making abstract themes tangible.
- **Materials Needed**: A small table, a magnifying glass,

a jar or tray, rotating objects from nature or loaned from a museum.

- **Making it Interactive**: Include a small journal where students can sketch their observations or write down "I wonder..." questions about the specimens.

5. Idea: "The Legend" - Key Vocabulary Word Wall

- **Type**: Functional & Organizational System
- **Core Concept**: A word wall designed as a "map legend." New vocabulary words are added on cards that look like map keys. Each word includes a simple definition and a visual symbol.
- **Why It's Important**: This provides crucial semantic support for all learners, especially English Language Learners. The thematic design integrates the vocabulary practice seamlessly into the classroom environment, making word acquisition a natural part of exploration.
- **Materials Needed**: Cardstock, velcro dots, markers, a large world map or parchment-style paper as a background.
- **Making it Interactive**: During writing or discussion, challenge students to "use a word from the legend." Students can suggest new words to add as they discover them in their reading.

The "Why" Behind This Strategy: This plan works because it moves decor from passive to active. The room becomes a toolkit for learning, not a museum of teacher-made posters. This student-centered, interactive approach reduces the teacher's workload in the long run (students help maintain it), increases engagement, and creates a space that is both intellectually stimulating and emotionally safe—a true basecamp for young explorers.

Bonus: Pro-Tip for Teachers - Design on a Dime:

- **Fabric Over Paper**: Use cheap fabric from a discount store as a bulletin board background. It doesn't fade, won't tear, and can be reused year after year.

- **Lighting is Everything**: Ditch the harsh fluorescents. A few inexpensive floor and desk lamps with warm-toned bulbs can transform the room's entire mood into a calm, focused "explorer's lodge."

- **Nature is Free**: Collect branches, stones, and pine-cones to create low-cost, textural displays. A "reading raft" can be defined by a large piece of blue fabric and a few river rocks.

Prompt: Classroom Reward System

Simple Prompt: "Act as a 'Behavior & Motivation Architect'. Your task is to create a balanced, research-based set of classroom reward system ideas for [GRADE LEVEL], designed to reinforce effort, positive behavior, and community building for [GRADE LEVEL]".

- **Tone**: Positive, Motivational
- **Format**: Reward System Framework
- **Platform**: Classroom Management Plans
- **AI Role**: Behavior & Motivation Architect
- **Prompt Goal**: Encourage good behavior and effort through rewards
- **Tags**: #classroomRewards, #studentMotivation, #engagement
- **Prompt Variables**: {grade/level}, {classSize}, {rewardTypes}

Best Practice Tip: Mix extrinsic rewards (stickers, points) with intrinsic motivators (responsibility roles, praise).

Enhanced Prompt:

Act as a 'Behavior & Motivation Architect'. Your task is to create a comprehensive menu of classroom reward system ideas for [GRADE LEVEL]. The goal is to generate strategies that are not just about giving "prizes," but are intentionally designed to foster intrinsic motivation, build a positive classroom community, reinforce specific behaviors, and teach students the value of delayed gratification and collective responsibility.

The guide must be structured as follows:

1 **The "Why" it Matters**: A brief, impactful explanation of the core philosophy behind effective reward systems (e.g., "An effective reward system is a teaching tool, not a bribery system. Its purpose is to intentionally recognize and celebrate desired behaviors, academic habits, and social skills, making them more likely to be repeated. The best systems focus on effort, growth, and community, shifting the classroom culture from compliance to shared accomplishment and intrinsic satisfaction.").

2 **Overarching Goal**: State the primary objective of this motivational system (e.g., "To create a predictable, positive environment where students feel seen and valued for their contributions, and where the focus is on cultivating proactive habits and a strong learning community, rather than simply managing misbehavior.").

3 **The Reward System Menu**: A Balanced Approach Provide a list of 5-7 distinct system ideas. Ensure there is a deliberate mix of the following types to cater to different motivational needs and classroom goals:

   ◇ **Individual Recognition Systems**: To celebrate personal achievement and effort.

   ◇ **Group/Whole-Class Systems**: To build teamwork and collective responsibility.

   ◇ **Privilege-Based Systems**: To offer experiences instead of material items.

   ◇ **Mystery & Gamified Systems**: To build excitement and engagement.

4 **Breakdown for Each System Idea**: For every idea, include:

◇ **System Name**: A catchy, descriptive title.

◇ **How It Works**: A concise, step-by-step explanation of how to set up and run the system.

◇ **Target Behaviors**: The specific habits or actions this system is designed to reinforce (e.g., "Completing homework," "Helping a peer," "Transitioning quietly.").

◇ **The "Why" it's Important**: A brief explanation of the psychological or social skill it promotes (e.g., "This system visualizes progress and teaches the value of working toward a long-term goal, fostering persistence and delayed gratification.").

5   **The "Why" Behind This Strategy**: A concluding summary explaining how this balanced, thoughtfully designed approach avoids the pitfalls of extrinsic reward systems (e.g., saturation, entitlement) and instead builds a self-sustaining culture of positivity, mutual respect, and internal motivation.

6   **Bonus**: Pro-Tip for Teachers - Implementation & Sustainability:

◇ **Focus on the "Why"**: Always explain why a behavior is being rewarded. Connect it to classroom values like respect, learning, and safety.

◇ **Keep it Fresh**: Rotate systems every quarter to maintain novelty and excitement.

◇ **Involve Students**: Have students brainstorm and vote on reward ideas. This creates immense buy-in.

◇ **Praise the Behavior, Not Just the Reward**:

> The most powerful reinforcement is specific, verbal praise. The reward should be a bonus, not the sole focus. Say, "I am so impressed by how you persevered through that challenging problem," not just "Here's a ticket."

The final output should be a complete, plug-and-play motivation toolkit that provides a clear framework for building a positive and self-regulated classroom community.

Complete Output:

The "Why" it Matters: Rewards are not about buying good behavior; they are about strategically highlighting it. A well-designed system makes the abstract concept of "doing the right thing" visible, tangible, and celebrated. It teaches students that their positive choices have positive consequences, for themselves and their community, which is a foundational life lesson.

The Reward System Menu: Strategies for a Motivated Classroom

1. System: "Bright Spot" Recognition

- **How It Works**: Keep a stack of unique, positive notes (e.g., "You Shined Today!"). When you catch a student demonstrating a target behavior, quietly place the note on their desk. Students collect them.

- **Target Behaviors**: Effort, kindness, perseverance, participation.

- **Why It's Important**: This provides immediate, private, and specific positive reinforcement without disrupting the lesson. It makes students feel personally seen and valued for their character, not just their academic performance.

2. System: Class "Marble Jar" or "Puzzle Pieces"

- **How It Works**: Have a clear jar and a bag of marbles (or a puzzle with a few pieces). Award marbles/pieces to the whole class for demonstrating collective target behaviors (e.g., everyone turning in homework, receiving a compliment from another teacher, a smooth transition). When the jar is full or the puzzle is complete, the class earns a reward.

- **Target Behaviors**: Whole-class routines, teamwork, positive behavior in specials.

- **Why It's Important**: This builds a sense of shared purpose and community. Students learn to encourage each other and work toward a common goal, understanding that their individual actions contribute to the group's success.

3. System: Reward Choice Coupons

- **How It Works**: Create a "menu" of free or low-cost privilege-based rewards (e.g., "Eat lunch with the teacher," "15 minutes of free tech time," "Swap seats with a friend for a day," "Wear a hat in class"). Students earn coupons for individual achievements that they can redeem.

- **Target Behaviors**: Academic milestones, achieving a personal goal.

- **Why It's Important**: This system is sustainable and budget-friendly. It values experiences over toys or candy, and allows students to choose a reward that is personally meaningful, increasing its motivational power.

4. System: "Mystery Motivator"

- **How It Works**: Write a secret reward on a piece of paper and seal it in an envelope. Draw a simple grid on the whiteboard. The class earns chances to reveal squares on the grid (e.g., for each homework submission, a square is revealed). When a pre-determined icon is uncovered, the class wins the mystery reward.

- **Target Behaviors**: Repetitive, whole-class behaviors like homework completion or hallway behavior.
- **Why It's Important**: The element of mystery and chance is highly engaging and builds anticipation. It leverages the power of variable reinforcement, which is a powerful motivator.

5. System: "Blurt Beans" (For Individual Self-Management)

- **How It Works**: Give each student a small container with 3-5 beans (or tokens) at the start of the day. The rule is that they must raise a quiet hand to speak. If they "blurt" out, they lose a bean. Students who have beans left at the end of the day get a small reward or enter a drawing.
- **Target Behaviors**: Self-regulation, impulse control, respectful listening.
- **Why It's Important**: This provides a concrete, visual tool for students to manage their own behavior. It's a non-verbal cue that helps them develop executive functioning skills without public shaming.

The "Why" Behind This Strategy: This plan works because it moves beyond a one-size-fits-all sticker chart. By offering a menu of systems, you can choose one that fits your class's specific needs and your teaching style. The focus on non-tangible, experience-based rewards and community building fosters intrinsic motivation and creates a positive classroom culture where students feel empowered and celebrated.

Bonus: Pro-Tip for Teachers - Implementation & Sustainability:

- **Phase Out Over Time**: As positive habits become ingrained, gradually phase out the tangible reward and shift to intermittent praise. The goal is for the behavior itself to become the norm.

- **Be Consistent**: The system only works if it is applied fairly and consistently. Set clear rules for how rewards are earned from the start.

- **Avoid Taking Away What's Been Earned**: Never remove a marble from the jar or a coupon once it's been earned. This destroys trust and undermines the entire positive system. Focus on adding, not subtracting.

Category 3: Assessment & Feedback

Theme

This category focuses on the critical processes of evaluation and feedback within schools. It includes a range of tools such as quizzes, exams, rubrics, surveys, and reflection activities. These resources are designed to support teachers in effectively measuring student learning and progress.

By utilizing these tools, educators are able to provide students with meaningful and constructive guidance. The emphasis is on not only assessing knowledge and skills but also on fostering a growth mindset through reflective practices and targeted feedback.

Category Goal

The purpose of implementing assessment frameworks and feedback systems is to provide educators with effective tools that support tracking student progress. These systems are designed to facilitate fair grading practices and to promote reflective learning among students.

By using structured assessment methods, teachers can ensure that evaluations are consistent and transparent, helping students understand their strengths and areas for improvement. Additionally, feedback mechanisms encourage students to engage in self-reflection, reinforcing a growth mindset and supporting ongoing academic development.

Mini-Index: Assessment and Feedback

Prompt	AI Role	Prompt Goal
Quiz & Assessment Builder	Assessment Designer/ Curriculum Alignment Specialist	Provide ready-to-use quizzes for knowledge checks
Rubric De-signer	Teacher / Assessment for Learning Specialist	Ensure transparent and consistent grading
Formative Assessment Generator	Formative Assessment Coach / Data-Informed Teaching Specialist	Track student understanding and adjust teaching
Summative Assessment Generator	Summative Assessment Architect / Exam De-signer	Evaluate comprehensive mastery of a unit or subject
Lesson Reflection Template	Reflective Practice Mentor / Instructional Growth Coach	Encourage students to self-assess their learning
Student Survey Generator	Student Feedback Architect / Instructional Improvement Coach	Collect feedback on class experiences
Classroom Assessment Techniques	Formative Assessment Coach / Instructional Strategist	Provide quick tools to check understanding
Feedback & Grading Rubric	Assessment for Learning Coach / Transparent Design Specialist	Combine fair grading with personalized feedback
Letter of Recommen-dation	Student Advocacy Specialist / Academic Mentor	Draft strong recommendation letters for students
Resume Enhancer (Academic)	Career Strategy Consultant / Academic Resume Specialist	Optimize student resumes for admissions or roles

Detailed Prompts

Prompt: Quiz & Assessment Builder

Simple Prompt: "Act as an 'Assessment Designer' and 'Curriculum Alignment Specialist'. Your task is to create a comprehensive teacher-ready, standards-aligned quiz on the topic of [TOPIC] for [GRADE LEVEL/COURSE]".

- **Tone**: Objective, Academic
- **Format**: Quiz (10 Questions + Answers)
- **Platform**: Classroom, LMS, Exam Papers
- **AI Role**: Assessment Designer/Curriculum Alignment Specialist
- **Prompt Goal**: Provide ready-to-use quizzes for knowledge checks and exams
- **Tags**: #quiz, #assessment, #evaluation
- **Prompt Variables**: {topic}, {questionTypes}, {difficultyLevel}

Best Practice Tip: Include a mix of question types (MCQ, short, problem-solving) to assess a variety of skills.

Enhanced Prompt:

Act as an 'Assessment Designer' and 'Curriculum Alignment Specialist'. Your task is to create a comprehensive, standards-aligned quiz on the topic of [TOPIC] for [GRADE LEVEL/COURSE]. The goal is to generate an assessment that effectively measures student mastery of core concepts, provides valuable data on student understanding, and is efficient for the teacher to grade and analyze.

The guide must be structured as follows:

1. **The "Why" it Matters**: A brief, impactful explanation of the core principle behind effective assessment design (e.g., "A well-designed quiz is not just an evaluation tool but a critical diagnostic instrument that informs future instruction by pinpointing student misconceptions and confirming mastery of key learning objectives.").

2. **Alignment to Learning Objectives**: List the 2-3 primary learning objectives or standards that this quiz is designed to assess (e.g., "1. Students will be able to define and identify key components of the water cycle. 2. Students will be able to explain the processes of evaporation, condensation, and precipitation.")

3. **The Quiz**: A Balanced Assessment Provide a set of [NUMBER] questions. The quiz must include a strategic mix:

    ◇ **Multiple-Choice Questions (MCQs)**: Designed to efficiently assess breadth of knowledge and identify common misconceptions.

    ◇ **Short-Answer Questions**: Designed to assess depth of understanding, reasoning, and ability to articulate knowledge.

4. **Breakdown for Each Question**: For every question, include:

    ◇ **Question Number & Text**: The question presented clearly.

    ◇ **Question Type**: Label as Multiple Choice or Short Answer.

    ◇ **Target Objective**: Specify which learning objective (from section 2) this question assesses.

◇ **Cognitive Level**: Categorize the question based on its demand (e.g., Recall, Application, Analysis).

◇ **Correct Answer**: For MCQs, provide the correct letter and a full answer statement. For short answer, provide a model answer.

◇ **For MCQs Only**: Distractor Analysis: Briefly explain the pedagogical reason for each incorrect answer (e.g., "Option B is a common misconception where students confuse [X] with [Y]").

5 **Grading Rubric for Short-Answer Questions**: Provide a simple, clear, and actionable single-point rubric for the short-answer questions.

◇ **Column 1**: Criteria for Success (What a perfect answer includes)

◇ **Column 2**: Room for Notes (For teacher comments on how the student's answer met or missed the criteria)

6 **The "Why" Behind This Strategy**: A concluding summary explaining how this balanced and objective-driven approach to assessment provides more reliable data on student learning, saves the teacher time in analysis, and directly informs whether to re-teach, move on, or provide enrichment.

7 **Bonus**: Pro-Tip for Data-Driven Instruction: Add a short section on how the teacher can quickly analyze the results after grading (e.g., "Tally the number of incorrect answers for each MCQ distractor. This will immediately show the most common class-wide misconceptions to address in the next lesson.").

The final output should be a complete, plug-and-play assessment kit that is easy to administer, grade, and use to guide future teaching decisions.

Complete Output

Grade Level: 7 | Subject: Science | Topic: The Water Cycle

1. The "Why" it Matters

A well-designed quiz is not just a way to grade students — it is a diagnostic tool. This assessment ensures students understand the water cycle not only at the recall level (definitions, stages) but also at deeper levels (application, analysis). It pinpoints misconceptions (e.g., mixing evaporation and condensation) and provides evidence for whether to re-teach, enrich, or move forward.

2. Alignment to Learning Objectives: By the end of this quiz, students should be able to:

- Identify and define the key stages of the water cycle (evaporation, condensation, precipitation, collection).
- Explain the processes of the water cycle using scientific reasoning.
- Analyze the importance of the water cycle in Earth's systems and ecosystems.

3. The Quiz: A Balanced Assessment (10 Questions)

Multiple Choice (Q1–Q6)

Q1. Which process changes liquid water into water vapor?

    A. Condensation

    B. Evaporation

    C. Precipitation

D. Collection

Type: MCQ | Target Objective: LO1 | Cognitive Level: Recall

Correct Answer: B. Evaporation

Distractor Analysis:

> A: Condensation = opposite process.
>
> C: Precipitation = water falling to Earth.
>
> D: Collection = water gathering in oceans/lakes.

Q2. Clouds form in the atmosphere because of which process?

> A. Precipitation
>
> B. Evaporation
>
> C. Condensation
>
> D. Collection

Type: MCQ | Target Objective: LO1 | Cognitive Level: Recall

Correct Answer: C. Condensation

Distractor Analysis:

> A: Precipitation happens after clouds form.
>
> B: Evaporation adds vapor but does not form clouds.
>
> D: Collection = after precipitation.

Q3. Which of the following is an example of precipitation?

> A. Water running into rivers
>
> B. Snow falling during winter
>
> C. Clouds forming in the sky
>
> D. Puddles drying on a hot day

Type: MCQ | Target Objective: LO1 | Cognitive Level: Recall

Correct Answer: B. Snow falling during winter

Distractor Analysis:

    A: Runoff = collection.

    C: Cloud formation = condensation.

    D: Drying = evaporation.

Q4. If the Sun's energy decreases significantly, which stage of the water cycle would be MOST affected?

    A. Evaporation

    B. Condensation

    C. Precipitation

    D. Collection

Type: MCQ | Target Objective: LO2 | Cognitive Level: Application

Correct Answer: A. Evaporation

Distractor Analysis:

- B & C depend on evaporation first.
- D = less directly affected.

Q5. Which stage returns water to lakes, oceans, and rivers?

    A. Precipitation

    B. Condensation

    C. Evaporation

    D. Collection

Type: MCQ | Target Objective: LO1 | Cognitive Level: Recall

Correct Answer: D. Collection

Distractor Analysis:

- A = falling of water, not storage.
- B & C = phase changes, not storage.

Q6. Why is the water cycle important for living organisms?

> A. It prevents air pollution.
>
> B. It provides clean water and regulates ecosystems.
>
> C. It creates energy for plants.
>
> D. It keeps rocks from breaking down.

Type: MCQ | Target Objective: LO3 | Cognitive Level: Analysis

Correct Answer: B. It provides clean water and regulates ecosystems.

Distractor Analysis:

- A & D = unrelated.
- C = energy for plants comes from sunlight, not the water cycle.
- Short Answer (Q7–Q10)

Q7. Define condensation in your own words and give an example.

- **Type**: Short Answer | Target Objective: LO1 | Cognitive Level: Recall & Application
- **Model Answer**: Condensation is when water vapor cools and changes back into liquid. Example: water droplets forming on the outside of a cold glass.
- Q8. Explain how evaporation and condensation work together to form clouds.

- **Type**: Short Answer | Target Objective: LO2 | Cognitive Level: Application
- **Model Answer**: Evaporation adds water vapor into the air. As the vapor rises and cools, condensation turns it into droplets that cluster together to form clouds.

Q9. Imagine a region stops receiving rainfall for several months. Which part(s) of the water cycle are disrupted, and what would happen to plants and animals?

- **Type**: Short Answer | Target Objective: LO3 | Cognitive Level: Analysis
- **Model Answer**: Precipitation and collection are disrupted. Rivers and lakes would shrink, plants would dry out, and animals would struggle to find water, possibly leading to ecosystem collapse.

Q10. Describe one way humans can impact the water cycle and explain its effect.

- **Type**: Short Answer | Target Objective: LO3 | Cognitive Level: Evaluation
- **Model Answer**: Cutting down forests reduces transpiration, leading to fewer clouds and less rainfall. This can cause droughts and harm ecosystems.

4. Grading Rubric for Short Answers

Criteria for Success	Teacher Notes
Defines concept/process accurately	
Uses correct scientific vocabulary	
Provides a clear example or explanation	
Demonstrates reasoning (not just recall)	

5. The "Why" Behind This Strategy: This assessment

uses both breadth (MCQs) and depth (short answers) to capture learning. MCQs highlight misconceptions and are quick to grade; short answers require reasoning and communication skills. Distractor analysis supports re-teaching by showing why students may have chosen the wrong option.

6. Bonus: Pro-Tip for Data-Driven Instruction: After grading:

- Tally incorrect responses for each MCQ distractor. If many students chose "condensation" for Q1, reteach the distinction between condensation and evaporation.

- For short answers, look for patterns in misconceptions (e.g., confusing precipitation with runoff).

- Use this data to plan a mini-review lesson targeting those weak points.

This is a plug-and-play quiz kit: teacher can print, use in Google Forms, or adapt into a gamified tool like Kahoot.

Prompt: Rubric Designer

Simple Prompt: "Act as an 'Teacher' and 'Assessment for Learning Specialist'. Your task is to create a clear, standards-aligned grading rubric for a [ASSIGNMENT TYPE - e.g., Persuasive Essay, Science Lab Report, Oral Presentation] designed for [GRADE LEVEL/COURSE]".

- **Tone**: Clear, Objective
- **Format**: Grading Rubric Table
- **Platform**: Assignments, LMS, Teacher Guides
- **AI Role**: Teacher / Assessment for Learning Specialist
- **Prompt Goal**: Ensure transparent and consistent grading across student work
- **Tags**: #rubric, #grading, #assessment
- **Prompt Variables**: {assignment/project}, {criteria}, {performanceLevels}

Best Practice Tip: Share the rubric with students before starting the assignment for clarity and fairness.

Enhanced Prompt:

Act as an 'Teacher' and 'Assessment for Learning Specialist'. Your task is to create comprehensive grading rubric for a [ASSIGNMENT TYPE - e.g., Persuasive Essay, Science Lab Report, Oral Presentation] designed for [GRADE LEVEL/COURSE]. The goal is to produce a rubric that is not just a scoring sheet but a clear roadmap for excellence, providing students with explicit criteria, actionable feedback, and a deep understanding of the learning targets before they even begin the assignment.

The guide must be structured as follows:

1   **The "Why" it Matters**: A brief, impactful explanation of the core principle behind effective rubrics (e.g., "A well-designed rubric demystifies assessment. It shifts the teacher's role from a sole judge to a coach by making expectations transparent and providing a clear path for improvement. For students, it becomes a powerful self-assessment tool that reduces anxiety, promotes ownership of learning, and turns feedback from a final judgment into a conversation about growth.").

2   **Overarching Learning Objectives**: State the 2-3 primary skills or standards this assignment and its rubric are designed to assess (e.g., "Skill: Construct a coherent argument supported by credible evidence. Skill: Integrate and cite sources effectively. Skill: Utilize conventions of standard English for clarity.").

3   **The Rubric**: A Single-Point Design (Recommended for Clarity)

   ◇   **Rationale for Format**: The single-point rubric (with only the proficient column described) is recommended as it avoids negative language for developing learners and provides cleaner space for targeted feedback on how the student can grow toward or beyond the standard.

   ◇   **The Rubric Structure**: Create a clear table with:

      ▪   **Criteria (3-5)**: The essential components of the assignment (e.g., Thesis & Focus, Evidence & Support, Organization, Language & Conventions).

      ▪   **Proficient (Meets Standard)**: A clear description of what meeting the standard looks like for each criterion. This is the target.

- **Areas for Growth (Left Column)**: Space to note where the student is still developing and what is needed to reach proficiency.

- **Areas of Excellence (Right Column)**: Space to note where the student has exceeded the proficiency standard.

4 **The "How-To"**: A Simple 3-Step Guide for Providing Feedback

◇ **Step 1**: Assess Holistically. Read through the entire assignment without making marks to get a general impression.

◇ **Step 2**: Criterion-by-Criterion Analysis. Use the rubric as a checklist. For each criterion, determine if the student is Developing, Proficient, or Excelling. Write one specific, actionable comment in the corresponding box (e.g., "Your thesis is clear but could be more arguable. Try starting with 'Although...'").

◇ **Step 3**: Global Feedback & Next Steps. Write one overall comment that summarizes strengths and one specific "next step" for future improvement. This connects the dots between the individual criteria.

5 **The "Why" Behind This Strategy**: A concluding summary explaining how this transparent, criteria-focused approach to assessment fosters a growth mindset, makes grading faster and more consistent, and aligns every aspect of the assignment—from instruction to feedback—with the intended learning outcomes.

6 Bonus: Pro-Tip for Teachers - The Rubric as a Teaching Tool:

◇ **Co-Create the Rubric**: When possible, create the rubric with your students. Analyze exemplars and non-exemplars together to define the criteria for proficiency. This creates immense buy-in and understanding.

◇ **Students Self-Assess**: Have students use the rubric to score their own work or a peer's work before turning it in. This is a powerful metacognitive exercise that dramatically improves final submission quality.

◇ **Focus Your Grading**: You do not need to comment on every single error. Use the rubric to focus your feedback on the 1-2 most important areas for growth for that particular student.

The final output should be a complete, plug-and-play assessment kit that includes a ready-to-use rubric and a clear methodology for providing feedback that students will actually use.

Complete Output

ASSIGNMENT TYPE: Persuasive Essay

The "Why" it Matters:

A rubric is a promise of transparency. It tells students, "This is exactly what I value, and this is what success looks like." It eliminates the "guess what's in the teacher's head" game and empowers students to evaluate their own work as they create it. This transforms the assignment from a task to be completed into a goal to be mastered.

Overarching Learning Objectives:

- Students will craft a persuasive claim and support it with logical reasoning and relevant evidence.
- Students will structure an argument in a clear, cohesive manner.
- Students will acknowledge and respond to counterarguments.

The Rubric: Single-Point Design

Areas for Growth (Feedback for next time)	Criteria	Proficient (Meets Standard)	Areas of Excellence (You went above & beyond!)
The argument is unclear or not debatable.	Claim & Focus	• Presents a clear, specific, and arguable thesis statement. • Maintains a consistent focus on the claim throughout the essay.	• Thesis is nuanced and compelling. • Focus is exceptionally sharp and persuasive.
• Evidence is irrelevant, insufficient, or not integrated. • Sources are not cited.	Evidence & Support	• Supports the claim with relevant and sufficient facts, examples, and details. • Smoothly integrates evidence (e.g., with signal phrases). • Cites sources appropriately.	• Evidence is sophisticated and from a variety of credible sources. • Skillfully analyzes and explains how the evidence supports the claim.

Areas for Growth (Feedback for next time)	Criteria	Proficient (Meets Standard)	Areas of Excellence (You went above & beyond!)
• Ideas are disorganized or illogical. • Paragraphs lack structure.	Organization	• Uses a logical organizational structure (e.g., intro, body, conclusion). • Body paragraphs have clear topic sentences and supporting details. • Uses effective transitions to connect ideas.	• Organization is sophisticated and enhances the argument's impact. • Transitions are seamless and purposeful.
• Does not address other perspectives. • Counter-argument is a "straw man."	Counter-argument	• Acknowledges a relevant counterargument. • Provides a logical and respectful rebuttal or concession.	• Effectively integrates multiple counterarguments into the flow of the essay. • Rebuttal is particularly insightful and strengthens the main claim.
• Errors in grammar, spelling, or punctuation distract the reader.	Conventions	• Uses standard English conventions of grammar and usage. • Errors are few and do not interfere with meaning.	• Language is precise, powerful, and stylistically advanced. • Writing is polished and error-free.

The "How-To": A Simple 3-Step Guide for Providing Feedback

1. **Read for Flow**: Read the essay once without your pen. Get a sense of the overall argument and organization.

2. **Score with the Rubric**: Go criterion by criterion. Check the "Proficient" box for each met standard. For any unmet or exceeded standards, write a brief, specific comment in the "Areas for Growth" or "Areas of Excellence" column. Reference the specific language of the rubric.

3. **Summarize**: At the bottom, write one thing the student did brilliantly related to the strongest criterion and one "Next Step" for improvement related to the weakest criterion. This ensures a balance of positive and constructive feedback.

The "Why" Behind This Strategy: This plan works because it makes assessment transparent, consistent, and constructive. The single-point format focuses your energy on providing personalized feedback for growth rather than justifying a point deduction. Students receive clear, actionable steps for improvement, which increases the likelihood they will learn from the assignment and apply that learning to the next one.

Bonus: Pro-Tip for Teachers - The Rubric as a Teaching Tool:

- **Peer Review Protocol**: Train students to use this rubric for peer editing. This dramatically improves the quality of feedback they give each other and helps them internalize the criteria for a good essay.

- **Live-Rating in Class**: Project an anonymous student paragraph (with permission) and have the class use the rubric to assess it together. This models the thought process of evaluation.

- **Revision Tool**: Require students to turn in their initial rubric self-assessment with their final draft, noting what they changed based on their own evaluation. This holds them accountable for the revision process.

Complete Output-2

Rubric Designer — Persuasive Essay (Grade 8 ELA)

1. The "Why" it Matters:

A rubric is more than a scoring guide; it is a roadmap for success. When shared with students before an assignment, it removes the guesswork of "what the teacher wants" and transforms grading into coaching. Students can self-assess, revise with purpose, and celebrate their strengths while targeting growth. Teachers gain consistency, fairness, and clarity in assessment.

2. Overarching Learning Objectives:

This persuasive essay rubric is designed to assess:

- **Argument & Focus**: Crafting a debatable thesis and maintaining a clear stance.
- **Evidence & Support**: Integrating credible sources and logical reasoning.
- **Organization & Coherence**: Structuring the essay with clarity and flow.
- **Counterargument & Rebuttal**: Acknowledging and responding to opposing viewpoints.
- **Language & Conventions**: Using grammar, vocabulary, and style effectively.

3. The Rubric: Single-Point Design

Areas for Growth	Criteria	Proficient (Meets Standard)	Areas of Excellence
Argument is unclear, too broad, or opinion-only.	Claim & Focus	Presents a clear, specific, and arguable thesis; maintains consistent focus throughout essay.	Thesis is sophisticated, nuanced, and engaging; focus is laser-sharp and compelling.
Evidence is weak, irrelevant, or poorly explained.	Evidence & Support	Supports claim with sufficient, relevant evidence; integrates sources with signal phrases; cites appropriately.	Evidence is rich, varied, and from highly credible sources; analysis of evidence is insightful and persuasive.
Paragraphs are disorganized or lack transitions.	Organization & Coherence	Essay follows a logical structure (intro, body, conclusion); clear topic sentences; effective transitions.	Organization enhances impact of argument; transitions are seamless and rhetorical.
Counterarguments ignored or oversimplified.	Counterargument & Rebuttal	Acknowledges at least one counterargument; responds with a logical rebuttal or concession.	Skillfully integrates multiple counterarguments; rebuttals are powerful and strengthen the main claim.

Frequent grammar errors; weak or inconsistent style.	Language & Conventions	Uses standard English grammar, spelling, and punctuation with minimal errors; style is clear and academic.	Writing is polished, precise, and stylistically strong; error-free and powerful use of language.

4. The "How-To": A Simple 3-Step Feedback Guide

- **Assess Holistically**: Skim the essay without notes to get a sense of flow and impact.

- **Criterion-by-Criterion**: Use the rubric to evaluate each area. Mark "Proficient" when met; leave targeted notes under Growth or Excellence.

- **Global Feedback**: Summarize with one strength ("Your evidence variety was excellent...") and one next step ("Work on smoother transitions between ideas...").

5. The "Why" Behind This Strategy

This rubric creates 'clarity and fairness' for students and teachers. Students gain a clear model of success, while teachers reduce subjectivity by grounding evaluation in transparent criteria. It builds a 'growth mindset': instead of "I got an 80%," students hear, "I met the standard for evidence, and next time I can improve on counterarguments."

6. Bonus: Pro-Tip for Teachers — Rubric as a Teaching Tool

- **Co-Create with Students**: Show sample essays and build the rubric together. Students gain buy-in and internalize expectations.

- **Use for Peer Review**: Before submitting, have students exchange essays and score each other with the rubric. This deepens understanding of the criteria.

- **Anchor Revision**: Require students to reflect on their rubric feedback and describe how they improved before resubmitting.

This plug-and-play rubric is ready to be printed, shared digitally, or uploaded into an LMS for seamless grading and feedback.

Prompt: Formative Assessment Generator

Simple Prompt: "Act as a 'Formative Assessment Coach' and 'Data-Informed Teaching Specialist'. Your task is to create a clear, classroom-ready guide to formative assessment strategies educators can use in real time".

- **Tone**: Supportive, Practical
- **Format**: Activity List (x5)
- **Platform**: Classroom, LMS
- **AI Role**: Formative Assessment Coach / Data-Informed Teaching Specialist
- **Prompt Goal**: Track student understanding and adjust teaching in real time
- **Tags**: #formativeAssessment, #studentFeedback, #evaluation
- **Prompt Variables**: {topic}, {grade/level}, {learning-Objectives}

Best Practice Tip: Use exit tickets or one-minute papers to quickly capture student comprehension.

Enhanced Prompt:

Act as a 'Formative Assessment Coach' and 'Data-Informed Teaching Specialist'. Your task is to create a comprehensive guide to formative assessment methods for educators. The goal is to provide a toolkit of low-effort, high-impact strategies that provide real-time evidence of student understanding, allowing teachers to adjust instruction during the learning process, not just at the end.

The guide must be structured as follows:

1. **The "Why" it Matters**: A brief, impactful explanation of the core principle behind formative assessment (e.g., "Formative assessment is the compass, not the autopsy. It is the ongoing process of gathering evidence of learning while teaching is still happening. This real-time feedback loop is the most powerful tool a teacher has to close the gap between what students currently understand and the desired learning goal, ensuring that no student is left behind due to misunderstood instruction.")

2. **Overarching Goal**: State the primary objective of using these methods (e.g., "To move from teaching on autopilot to teaching with intention. These methods provide a constant pulse on student understanding, enabling teachers to make informed, in-the-moment decisions about whether to re-teach, pivot, or move on, thereby maximizing the effectiveness of every instructional minute.").

3. **The Formative Assessment Toolkit**: A Strategic Mix Provide a list of 5-7 distinct methods. The methods must be categorized by their primary function and data output:

   ◇ **Quick Checks for Understanding**: To gauge whole-class comprehension in seconds.

   ◇ **Promoting Metacognition**: To get students to think about their own thinking.

   ◇ **Encouraging Discussion & Dialogue**: To assess understanding through conversation.

4. **Breakdown for Each Assessment Method**: For every method, include:

   ◇ **Method Name**: A clear, memorable name.

◇ **Category**: Label it as Quick Check, Metacognition, or Discussion.

◇ **How It Works**: A concise, step-by-step explanation of how to implement it in a live classroom setting.

◇ **The "Why" it's Important**: A brief explanation of the specific cognitive or pedagogical benefit (e.g., "This method provides anonymous, whole-class data instantly, allowing the teacher to see if one student is confused or if the entire class missed the concept, without putting any individual student on the spot.").

◇ **What to Do with the Data**: The critical next step—how a teacher should respond to the information gathered.

5 **The "Why" Behind This Strategy**: A concluding summary explaining how this ongoing, low-stakes approach to assessment reduces student anxiety, creates a culture of mistake-making and growth, and transforms teaching from a one-size-fits-all lecture into a responsive, differentiated practice.

6 **Bonus**: Pro-Tip for Teachers - The Cycle of Inquiry:

◇ **The Loop**: Remember the cycle: Teach –> Assess (using a method) -> Analyze the data -> Adjust -> Re-teach (if needed). The assessment is useless without the action that follows.

◇ **Start Small**: Choose one method to use once per day for a week. Master it before adding another.

◇ **No-Grade Zone**: The power of formative assessment is negated if it's graded. Keep it low-stakes to encourage honest responses.

◇ **Think Beyond Thumbs**: While thumbs up/down is a start, strive for methods that provide more nuanced data (e.g., why a student is confused).

The final output should be a complete, plug-and-play assessment toolkit that empowers educators to become more responsive and data-informed in their teaching practice.

Complete Output:

The "Why" it Matters: Summative assessments (tests, quizzes) diagnose what students learned after the fact. Formative assessment is the ongoing monitoring of the learning process. It's the difference between a doctor performing an annual physical (summative) and a doctor continuously monitoring a patient's vitals during surgery (formative). The latter allows for immediate intervention, which is precisely what effective teaching requires.

The Formative Assessment Toolkit: Your Pulse on Student Learning

1. Method: The Exit Ticket

- **Category**: Quick Check for Understanding
- **How It Works**: In the last 3-5 minutes of class, students answer a single question or solve one problem on a small slip of paper. It must be turned in as they "exit." The prompt can be: "What was the most important thing you learned today?" "What question do you still have?" or a direct problem to solve.
- **Why It's Important**: It provides a perfect snapshot of understanding at the end of a lesson. It forces students to retrieve and synthesize information, and it gives the

teacher a crucial planning tool for the next day's lesson.

- **What to Do with the Data**: Sort the tickets into three piles: "Got it," "Almost there," "Not there." Use this to form strategic groups for the next day's warm-up or to decide if the whole class needs a re-teach.

2. Method: Four Corners

- **Category**: Quick Check for Understanding/Discussion
- **How It Works**: Pose a question with four possible answers (A, B, C, D) or four levels of agreement (Strongly Agree, Agree, Disagree, Strongly Disagree). Assign each answer a corner of the room. Students physically move to the corner that represents their answer.
- **Why It's Important**: It incorporates movement (kinesthetic learning) and instantly visualizes the distribution of understanding or opinion across the entire class. It's a great way to spark debate and discussion.
- **What to Do with the Data**: Ask students in different corners to defend their reasoning. This allows students to teach each other. If most of the class is in the wrong corner, it's a clear signal for immediate whole-class re-teaching.

3. Method: Think-Pair-Share

- **Category**: Discussion / Metacognition
- **How It Works**: Pose a challenging, open-ended question. First, give students individual time to Think and jot down ideas. Then, have them Pair up with a partner to discuss their thoughts. Finally, Share their or their partner's ideas with the whole class.
- **Why It's Important**: It ensures all students have time to process and formulate an answer, not just the quickest hand-raisers. It builds language skills and allows for low-risk practice before sharing with the entire class.

- **What to Do with the Data**: Circulate during the "Pair" phase to listen in on conversations. This is where you hear the real misconceptions and brilliant ideas. Use what you hear to select which pairs will share and to guide the whole-class discussion.

4. Method: One-Minute Paper

- **Category**: Metacognition
- **How It Works**: Mid-lesson, pause and give students exactly one minute to answer a targeted prompt: "What is the clearest point so far?" "What is the muddiest point so far?" or "How does this connect to what we learned yesterday?"
- **Why It's Important**: It provides a "check-in" during the learning process, not just at the end. It teaches students to monitor their own understanding in real-time (metacognition) and gives you a chance to clarify confusion before moving on to the next concept.
- **What to Do with the Data**: Quickly scan the responses. If you see a common "muddy point," stop and address it immediately. This prevents small confusions from snowballing into major misunderstandings.

5. Method: Mini Whiteboards

- **Category**: Quick Check for Understanding
- **How It Works**: Every student has a small whiteboard, marker, and eraser. Pose a question. All students write their answer and hold it up on your count.
- **Why It's Important**: It guarantees 100% participation and provides immediate, visible data from every single student. It's perfect for quick math problems, vocabulary, short answers, or multiple-choice questions.
- **What to Do with the Data**: Do a quick visual scan. If you see a wide variety of answers, have students turn

and convince their neighbor they are right, then revote. This promotes academic discourse and self-correction.

The "Why" Behind This Strategy: This plan works because it treats assessment as a teaching tool, not a grading tool. These methods are fast, low-stakes, and embedded directly into instruction. They shift the classroom culture from "Did you get the right answer?" to "How are we all thinking about this?" This creates a safe environment for risk-taking and provides the teacher with the actionable data needed to be truly responsive to student needs.

Bonus: Pro-Tip for Teachers - The Cycle of Inquiry:

- **The Most Important Step**: The assessment itself is pointless without the instructional adjustment. Always ask yourself: "Based on what I just learned, what will I do next?" This is the core of responsive teaching.

- **Variety is Key**: Don't use the same method every day. Rotate them to keep students engaged and to assess different types of thinking.

- **Involve Students**: Sometimes, after using a method, share the aggregated data with the class. Say, "I noticed about half of us are confident with this, and half are still unsure. Let's pair up so you can teach each other." This builds a collaborative learning community.

Complete Output -2

Grade Level: 6 | Subject: Math | Topic: Fractions

1. The "Why" it Matters

Formative assessment is teaching with a compass in hand. Instead of waiting until a test to discover misconceptions, it gives teachers real-time insight into how students are thinking. For a concept like fractions—which is abstract, visual, and often misunderstood—formative assessment ensures gaps are caught early, helping every learner build a solid foundation.

2. Overarching Goal

To provide teachers with low-prep, high-impact strategies to monitor student understanding of fractions during instruction, adjust lessons in the moment, and empower students to reflect on their learning process.

3. The Formative Assessment Toolkit

1. Method: Fraction Quick Draw

- **Category**: Quick Check for Understanding
- **How It Works**: Ask students to draw a model of a fraction (e.g., 3/4) on a mini whiteboard—circle, rectangle, or number line. They hold it up simultaneously.
- **Why It's Important**: Instantly reveals whether students understand part-whole representation or confuse numerator and denominator.
- **What to Do with the Data**: If many draw incorrectly, pause for reteaching with manipulatives. If only a few, pair them with peers who got it right for peer explanation.

2. Method: One-Minute Paper — "What's Tricky?"

- **Category**: Metacognition
- **How It Works**: Halfway through the lesson, students jot down: "The clearest part so far is…" and "The trickiest part so far is…".
- **Why It's Important**: Promotes self-awareness and gives you a window into which steps in fractional operations (like finding common denominators) are causing

confusion.

- **What to Do with the Data**: Collect and skim—if multiple students' flag "adding unlike denominators," plan a guided example immediately before moving forward.

3. Method: Fraction Corners

- **Category**: Discussion / Quick Check
- **How It Works**: Post four answer options (A, B, C, D) around the room to a fraction comparison question (e.g., Which is larger: 3/5 or 2/3?). Students move to the corner that matches their choice.
- **Why It's Important**: Builds movement and energy, while instantly visualizing misconceptions (like comparing denominators only).
- **What to Do with the Data**: Facilitate peer-to-peer explanation between corners. If one corner has many students with a misconception, stop and reteach that strategy.

4. Method: Think-Pair-Share — Fraction Word Problem

- **Category**: Discussion / Metacognition
- **How It Works**: Pose a real-world fraction problem (e.g., "You ate 3/8 of a pizza and your friend ate 2/8. How much is left?"). Students think silently, pair with a partner to compare reasoning, then share with the class.
- **Why It's Important**: Provides all students with a voice and highlights different solution strategies (visual models, subtraction, reasoning).
- **What to Do with the Data**: As you circulate, listen for strong strategies to highlight during the share. Identify if errors are conceptual (misunderstanding fractions) or procedural (calculation mistakes).

5. Method: Traffic Light Cards

- **Category**: Quick Check for Understanding
- **How It Works**: Each student has 3 colored cards (green = I get it, yellow = I'm unsure, red = I'm lost). During practice, ask students to hold up a card to show their current confidence.
- **Why It's Important**: Provides an anonymous, nonverbal way for students to signal understanding, reducing stigma.
- **What to Do with the Data**: If many hold yellow/red, slow down or reteach. If just a few, pull them into a small group during independent practice.

4. The "Why" Behind This Strategy

This toolkit ensures fractions don't become a silent struggle. By embedding simple checks into instruction, teachers gain immediate clarity on student thinking. Students feel safe to express confusion, teachers can pivot instruction in the moment, and learning becomes a responsive, collaborative process.

5. Bonus: Pro-Tip for Teachers — The Cycle of Inquiry

- Teach → Assess → Analyze → Adjust → Re-Teach is the golden loop of responsive instruction.
- **Start Small**: Choose one method (e.g., Exit Ticket or Quick Draw) to use daily for a week.
- **Keep it Low-Stakes**: Formative assessment should never be graded—its power lies in honest feedback.
- **Act on It**: The value of formative assessment is not in the data itself, but in what you do next.

This sample is ready-to-use for a Grade 6 Fractions lesson, but the same framework applies across any subject or grade.

Prompt: Summative Assessment Generator

Simple Prompt: "Act as a 'Summative Assessment Architect' and 'Exam Designer'. Your task is to create a comprehensive, classroom-ready menu of summative assessment formats and tasks for educators."

- **Tone**: Formal, Academic
- **Format**: Exam Paper / Summative Test
- **Platform**: Classroom, Standardized Testing
- **AI Role**: Summative Assessment Architect / Exam Designer
- **Prompt Goal**: Evaluate comprehensive mastery of a unit or subject
- **Tags**: #summativeAssessment, #exam, #evaluation
- **Prompt Variables**: {topic}, {grade/level}, {examDuration}

Best Practice Tip: Balance lower-level recall questions with higher-order analysis to cover Bloom's taxonomy.

Enhanced Prompt:

Act as a 'Summative Assessment Architect' and 'Exam Designer'. Your task is to create a comprehensive menu of summative assessment ideas for educators. The goal is to provide a range of options that move beyond traditional tests to authentically measure student mastery of standards, while also allowing for student choice, creativity, and the application of knowledge in novel contexts.

The guide must be structured as follows:

1 **The "Why" it Matters**: A brief, impactful explanation of the core principle behind effective summative assessment (e.g., "A summative assessment is the capstone of a learning unit. Its purpose is to evaluate student mastery of the key standards and objectives after instruction is complete. The best summative assessments are not just tests of memorization but are authentic performances of understanding that require students to synthesize knowledge, apply skills, and demonstrate their learning in a meaningful way.").

2 **Overarching Goal**: State the primary objective of summative assessment (e.g., "To provide a valid and reliable measure of student achievement against the defined learning targets, offering a clear snapshot of what a student knows and can do at the culmination of a period of learning. This data informs grades, placement, and provides evidence for program evaluation.").

3 **The Assessment Menu**: A Spectrum of Authentic Evidence Provide a list of 5-7 distinct assessment ideas. Ensure there is a deliberate mix of the following types to cater to diverse learning styles and measure different kinds of understanding:

   ◇ **Traditional & Standardized**: Reliable methods for measuring core knowledge and skills.

   ◇ **Performance-Based & Authentic**: Assessments that mirror real-world tasks and challenges.

   ◇ **Project-Based & Creative**: Long-term endeavors that allow for deep exploration and innovation.

   ◇ **Oral & Presentation-Based**: Assessments

that focus on communication and explanation.

4 **Breakdown for Each Assessment Idea**: For every idea, include:

◇ **Idea Name & Type**: A descriptive title and category (e.g., "Cumulative Portfolio (Project-Based)").

◇ **Core Concept**: A 1-2 sentence description of the assessment task.

◇ **The "Why" it's Important**: A brief explanation of what it uniquely measures and its long-term benefit (e.g., "This assesses a student's growth over time and their ability to curate, reflect on, and present their best work, fostering metacognition and pride in their learning journey.").

◇ **Key Components**: The essential elements required for the assessment (e.g., "A collection of 3-5 best works, a written reflection for each piece, a final summary statement.").

◇ **Ideal For**: The specific type of learning it best assesses (e.g., "Ideal for measuring growth, process, and mastery in writing, art, or design courses.").

5 **The "Why" Behind This Strategy**: A concluding summary explaining how offering a variety of summative assessment types provides a completer and more equitable picture of student learning. It allows different learners to shine, reduces the high-stakes pressure of a single test, and ensures that assessment is a valid demonstration of learning, not just test-taking skill.

6 **Bonus**: Pro-Tip for Teachers - Ensuring Validity and

Reliability:

◇ **Use a Rubric**: Every summative assessment, especially non-traditional ones, must be evaluated with a clear, criteria-based rubric shared with students in advance.

◇ **Backward Design**: The assessment should be designed after the learning objectives are set, not the other way around. Everything taught should directly build the skills needed for the summative task.

◇ **Offer Choice**: Where possible, provide students with a menu of 2-3 summative options to demonstrate their mastery. This increases buy-in and allows them to leverage their strengths.

◇ **Test the Test**: Always complete the assessment yourself first to check for clarity, timing, and alignment to the taught material.

The final output should be a complete, plug-and-play summative assessment guide that provides clear, rigorous, and fair ways to measure student learning at the end of a unit or course.

Complete Output:

The "Why" it Matters: Summative assessment is the "final proof" of learning. While formative assessment guides the journey, summative assessment confirms the destination. A well-designed summative task ensures that the time spent on a unit culminates in a meaningful demonstration of understanding, providing essential data for the student, teacher, and school on the effectiveness of instruction and the level of mastery achieved.

The Summative Assessment Menu: Measuring Mastery

1. Idea: Traditional Cumulative Exam

- **Type**: Traditional & Standardized
- **Core Concept**: A comprehensive written test covering the major concepts, skills, and vocabulary from the entire unit or semester.
- **Why It's Important**: It efficiently measures broad knowledge retention and the ability to recall and apply information under constraints. It prepares students for the format of standardized tests they will encounter.
- **Key Components**: A mix of question types: multiple-choice, short answer, matching, and essay.
- **Ideal For**: Efficiently assessing a large body of knowledge in subjects like History, Science, or Foreign Language vocabulary/grammar.

2. Idea: Performance Task or Simulation

- **Type**: Performance-Based & Authentic
- **Core Concept**: Students demonstrate their learning by performing a real-world task that requires application of their knowledge and skills (e.g., conducting a specific lab experiment and writing a formal report, debating a historical issue from a assigned perspective, solving a complex engineering design challenge).
- **Why It's Important**: It assesses the ability to use knowledge in a practical context, measuring critical thinking, problem-solving, and application skills that a paper test cannot.
- **Key Components**: A clear scenario, specific role for the student, defined deliverables (e.g., lab report, debate speech, design blueprint).
- **Ideal For**: STEM subjects, civics, business, and world languages (e.g., a simulated conversation).

3. Idea: Cumulative Portfolio

- **Type**: Project-Based & Creative
- **Core Concept**: A curated collection of a student's work over a unit or semester, accompanied by written reflections that explain the significance of each piece and demonstrate growth.
- **Why It's Important**: This assesses a student's meta-cognitive ability to evaluate their own work, track their progress, and articulate their learning journey. It values process over just the final product.
- **Key Components**: 3-5 selected pieces of work, a reflection for each piece, and a final introduction or summary statement.
- **Ideal For**: Writing courses, art, design, portfolio-based classes, and for showcasing long-term project development.

4. Idea: Oral Presentation or Defense

- **Type**: Oral & Presentation-Based
- **Core Concept**: Students present their findings, research, or analysis to an audience (teacher, peers, or a panel) and then answer questions about their work.
- **Why It's Important**: It assesses depth of understanding, communication skills, poise, and the ability to think on one's feet. Defending one's work requires a profound mastery of the content.
- **Key Components**: A presentation aid (slides, poster), a structured speech, and a Q&A session.
- **Ideal For**: Research projects, literary analysis, science fair projects, and capstone courses.

5. Idea: Synthesis Essay or Research Paper

- **Type**: Traditional & Standardized (with a focus on

depth)

- **Core Concept**: A formal written piece that requires students to synthesize information from multiple sources within the unit to form a unique argument or analysis.

- **Why It's Important**: It assesses higher-order thinking skills like analysis, evaluation, and synthesis, alongside writing proficiency. It shows if a student can connect ideas and construct a complex, evidence-based argument.

- **Key Components**: A clear thesis, evidence from multiple sources, proper citation, and a coherent structure.

- **Ideal For**: English Language Arts, social studies, and any humanities course where constructing arguments is key.

The "Why" Behind This Strategy: This plan works because it rejects a one-size-fits-all approach to final evaluation. By providing a menu of options, you acknowledge that students can demonstrate mastery in different ways. This leads to more accurate assessments. A creatively gifted student might thrive on a portfolio, while a logically minded student might prefer a traditional exam. This variety increases equity, reduces anxiety, and provides a much richer, more nuanced picture of what your students have truly learned.

Bonus: Pro-Tip for Teachers - Ensuring Validity and Reliability:

- **Blueprints are Key**: For exams, create an assessment blueprint that maps each question to a specific learning objective. This ensures your test actually measures what you taught.

- **Benchmark with Exemplars**: Before students begin a project or essay, show them examples of high, medium, and low-quality work. This makes your expectations tangible and the rubric easier to understand.

- **Blind Scoring**: When grading essays or projects, grade

all students' Question 1 first, then all Question 2, etc. This improves scoring consistency and reduces the "halo effect."

Complete Ouput-2

Summative Assessment Generator

Grade Level: 8 | Subject: Science | Topic: Forces & Motion

1. The "Why" it Matters: Summative assessments are the culmination of a learning journey. They allow both teacher and student to see whether the essential standards have been mastered. In science, this means not just recalling Newton's laws but applying them to real-world contexts. An effective summative assessment blends rigor with authenticity, ensuring students demonstrate their knowledge through reasoning, application, and communication.

2. Overarching Goal: To evaluate comprehensive mastery of forces and motion concepts—including laws of motion, balanced/unbalanced forces, acceleration, and real-world applications—using a mix of traditional, performance-based, and creative assessment formats.

3. The Summative Assessment Menu

1. Traditional Unit Exam

- **Type**: Traditional & Standardized
- **Core Concept**: A 50-minute written test with multiple-choice, short-answer, and one extended response.
- **Why It's Important**: Efficiently measures broad conceptual knowledge and procedural fluency. Ensures cov-

erage of the entire unit.

- **Key Components**:

    ◇ 15 MCQs (laws of motion, definitions, calculations)

    ◇ 5 short-answer questions (interpret graphs, explain force interactions)

    ◇ 1 extended response (apply Newton's 3rd law to a sports scenario)

- **Ideal For**: Measuring retention and breadth of knowledge across the unit.

2. Performance Task: Lab Investigation & Report

- **Type**: Performance-Based & Authentic
- **Core Concept**: Students design and conduct an experiment testing one of Newton's laws (e.g., rolling objects down ramps at different inclines).
- **Why It's Important**: Demonstrates ability to apply theory, collect data, analyze results, and draw evidence-based conclusions.
- **Key Components**: Experimental design, data table, graph, analysis paragraph, conclusion linked to Newton's laws.
- **Ideal For**: Assessing inquiry skills, critical thinking, and scientific reasoning.

3. Cumulative Portfolio — "Physics in My World"

- **Type**: Project-Based & Creative
- **Core Concept**: A collection of 3–5 artifacts (photos, diagrams, journal entries) documenting examples of forces and motion in daily life (e.g., seat belts, skateboarding, roller coasters).

- **Why It's Important**: Encourages transfer of classroom knowledge to authentic contexts, fostering long-term retention and relevance.

- **Key Components**: Artifacts with captions, one reflection per artifact, and a synthesis essay tying examples to Newton's laws.

- **Ideal For**: Students who excel in connecting science to real-world phenomena.

4. Oral Defense: "Physics in Action" Presentation

- **Type**: Oral & Presentation-Based

- **Core Concept**: A 5–7 minute oral presentation explaining how Newton's laws apply to a chosen real-world system (e.g., roller coasters, space rockets, bicycles). Includes Q&A from teacher/peers.

- **Why It's Important**: Builds communication and reasoning skills. Requires students to defend their understanding under questioning, ensuring deep mastery.

- **Key Components**: Slides or poster, prepared explanation, clear use of vocabulary, responses to at least two audience questions.

- **Ideal For**: Assessing depth of conceptual knowledge and oral communication.

5. Research Paper: "The Science Behind Sports"

- **Type**: Synthesis Essay / Research Paper

- **Core Concept**: Students research how forces and motion apply to a sport of their choice, citing at least three credible sources.

- **Why It's Important**: Assesses synthesis of multiple sources, ability to explain complex science in writing, and real-world application.

- **Key Components**: Thesis statement, structured argument, supporting evidence, proper citation, conclusion.

- **Ideal For**: Cross-curricular integration with literacy standards.

4. The "Why" Behind This Strategy

This menu ensures multiple pathways to success. Not all students excel in timed exams, but may thrive in labs, portfolios, or presentations. Offering varied summative options provides a fuller, more equitable picture of student mastery. It reduces test anxiety, validates diverse learning strengths, and ensures the assessment reflects authentic science practice.

5. Bonus: Pro-Tip for Teachers — Ensuring Validity & Reliability

- **Blueprint Alignment**: Create an assessment map showing how each task links directly to learning objectives (e.g., "Apply Newton's 2nd Law").

- **Use Rubrics**: For labs, portfolios, and essays, provide clear rubrics in advance to ensure fairness and transparency.

- **Balance Rigor**: Include both lower-level recall and higher-order tasks (application, synthesis).

- **Offer Choice**: Allow students to select between, for example, the oral defense or the research paper. This increases engagement and ownership.

- **Test the Test**: Complete a sample yourself to ensure timing and clarity are appropriate.

This output is ready-to-use for a Grade 8 Science: Forces & Motion summative assessment kit.

Prompt: Lesson Reflection Template

Simple Prompt: "Act as a 'Reflective Practice Mentor' and 'Instructional Growth Coach'. Your task is to create a comprehensive, classroom-ready reflection template for teachers to analyze the effectiveness of a [LESSON/TOPIC] taught to [GRADE LEVEL]."

- **Tone**: Reflective, Supportive
- **Format**: Reflection Worksheet / Journal Template
- **Platform**: Classroom, Student Journals
- **AI Role**: Reflective Practice Mentor / Instructional Growth Coach
- **Prompt Goal**: Encourage students to reflect on lessons and self-assess progress
- **Tags**: #reflection, #studentFeedback, #selfAssessment
- **Prompt Variables**: {lesson/topic}, {grade/level}

Best Practice Tip: Include both short-response and creative prompts (drawings, diagrams) for inclusiveness.

Enhanced Prompt:

Act as a 'Reflective Practice Mentor' and 'Instructional Growth Coach'. Your task is to create a comprehensive lesson reflection template for a teacher to use after teaching [LESSON/TOPIC] to [GRADE LEVEL]. The goal is to provide a structured yet flexible framework that moves beyond simple description to deep, analytical, and actionable reflection, transforming a single teaching experience into a personalized professional development opportunity.

The guide must be structured as follows:

1   **The "Why" it Matters**: A brief, impactful explanation of the core principle behind reflective practice (e.g., "Reflection is what transforms experience into expertise. A lesson taught without reflection is just a task completed; a lesson reflected upon is a dataset for growth. This process of critical self-inquiry is the most powerful, personalized, and readily available professional development tool a teacher has. It cultivates a growth mindset, builds self-efficacy, and is the engine of continuous instructional improvement.").

2   **Overarching Goal**: State the primary objective of this reflective practice (e.g., "To systematically analyze the cause-and-effect relationships within a lesson—between my instructional choices, student engagement, and learning outcomes—in order to identify a few specific, high-leverage adjustments that will increase my effectiveness in future lessons.").

3   **The Reflection Template**: A Structured Framework for Growth

  ◇ **Part 1**: The Facts (Objective Data): Quick, factual questions to set the context.

  ◇ **Part 2**: The Analysis (What Worked & Why): Questions designed to identify successes and connect them to specific teacher actions.

  ◇ **Part 3**: The Inquiry (Challenges & Root Causes): Questions that foster curiosity about difficulties and push beyond surface-level observations to underlying causes.

  ◇ **Part 4**: The Action (Synthesis & Forward Planning): The most critical section: translating insights into a concrete plan for the next lesson.

4 The "How-To": A Simple 3 Step Guide for Effective Reflection

◇ **Step 1**: Jot Down the "What" Immediately. Right after the lesson, take 2 minutes to quickly fill out "The Facts" and any burning impressions in the other sections. Capture the raw data before you forget.

◇ **Step 2**: Analyze the "Why" Later. Set aside 10-15 minutes later in the day to analyze your notes. Ask yourself "Why?" repeatedly. Why did that strategy work? Why did that student struggle? This is where the deep learning happens.

◇ **Step 3**: Commit to One "Next Step". From your analysis, choose the one most important adjustment you will make. This makes the reflection actionable and prevents it from being an overwhelming list of things to fix.

5 The "Why" Behind This Strategy: A concluding summary explaining how this structured approach to reflection prevents the common pitfalls of vague or overly critical self-assessment. It provides a balanced lens for viewing the lesson, forces evidence-based analysis, and ensures that the time spent reflecting directly translates into improved teaching practice.

6 Bonus: Pro-Tip for Teachers - Making Reflection a Habit:

◇ **Schedule It**: Block out 15 minutes in your calendar after teaching a key lesson. Treat this appointment with yourself as non-negotiable.

◇ **Be Kind and Specific**: Avoid general judgments like "that was bad." Instead, use evidence: "During the group work, three groups were off-task because the instructions were ambiguous." This is data, not failure.

◇ **Focus on One Thing**: You cannot fix every-thing at once. The goal of each reflection is to identify ONE high-impact change for next time. This makes growth manageable and sustainable.

◇ **Look for Patterns**: After several reflections, look back. Are you noticing a recurring chal-lenge? This pattern indicates an area for broader professional growth or coaching.

The final output should be a complete, plug-and-play reflection kit that empowers a teacher to turn every lesson, whether a tri-umph or a challenge, into a stepping stone for mastery.

Complete Output:

The "Why" it Matters: Teaching is a complex art built on thousands of micro-decisions. Reflection is the practice that brings those decisions into the light, allowing us to understand their impact. It is the difference between having 20 years of ex-perience and having one year of experience repeated 20 times. It is the deliberate practice that separates good teachers from truly great ones.

The Reflection Template: From Experience to Expertise

Lesson Name/Topic:_____ Date:_____Grade Level:_____

Part 1: The Facts (The Objective Picture)
- What was the planned learning objective?
- What were the main activities? (List in order)
- What formative assessment data did I gather? (e.g., exit ticket results, observation notes)

Part 2: The Analysis (Celebrating Success & Understanding Why)

- **What went exceptionally well? (Be specific**: "The turn-and-talk discussion on X was vibrant and productive.")

- What instructional choice or material likely contributed to this success? (e.g., "Providing a sentence stem gave all students access to the discussion.")

- Which students were highly engaged? When? Why might that be?

Part 3: The Inquiry (Investigating Challenges with Curiosity)

- Was there a moment that felt awkward or where the energy dropped? What was happening?

- Which students were disengaged or struggled? What were they struggling with? (e.g., "A small group struggled with the independent practice because they couldn't recall the first step of the process.")

- If I could re-teach this lesson, what is one thing I would change? Why?

- What surprised me?

Part 4: The Action (The Takeaway)

- Based on this reflection, what is the ONE most important thing I will do differently next time? (e.g., "I will add a visual cue of the problem-solving steps on the anchor chart for quick reference during independent work.")

- What is a lingering question I have? (e.g., "How can I better differentiate the reading for my struggling readers?")

- What will I carry forward into my next lesson?

The "How-To": A Simple 3-Step Guide for Effective Re-

flection

- **The 2-Minute Brain Dump (Right After Class)**: Quickly fill in Parts 1 and 3 with your immediate impressions. What sticks out? What felt off? Don't overthink it.

- **The 15-Minute Deep Dive (Later That Day)**: Revisit your template. Analyze your notes. For the successes in Part 2, identify your role. For the challenges in Part 3, ask "why" five times to get to the root cause.

- **The 1-Thing Promise (Before You Plan Next Time)**: Review your analysis and commit to the one actionable change in Part 4. This is your goal for your next planning session.

The "Why" Behind This Strategy: This plan works because it structures reflection as a constructive, data-informed conversation with yourself, not a critique session. By breaking it into factual, analytical, and actionable parts, it prevents overwhelm and ensures that your insights directly inform your future practice. This turns reflection from a nostalgic look back into a strategic plan forward, creating a powerful cycle of continuous improvement.

Bonus: Pro-Tip for Teachers - Making Reflection a Habit:

- **Use a Voice Memo**: If you're short on time, talk through the template questions into your phone's voice memo app on your drive home.

- **Reflect with a Partner**: Sometimes, do this with a colleague. Share your reflections and ask each other questions. An outside perspective can be invaluable.

- **Keep It All in One Place**: Maintain a reflection journal or a digital folder. Reviewing these over time is the best way to see your own professional growth and identify long-term trends.

Prompt: Student Survey Generator

Sample Prompt: "Act as a 'Student Feedback Architect' and 'Instructional Improvement Coach'. Your task is to create a research-informed, classroom-ready student survey for [SUBJECT/GRADE LEVEL] to be administered [e.g., mid-year, end of unit, end of semester] that gathers both quantitative and qualitative feedback".

- **Tone**: Neutral, Respectful
- **Format**: Survey (10 Questions)
- **Platform**: Google Forms, LMS, Paper Surveys
- **AI Role**: Student Feedback Architect / Instructional Improvement Coach
- **Prompt Goal**: Gather student feedback for teaching improvement
- **Tags**: #survey, #studentFeedback, #assessment
- **Prompt Variables**: {topic/classExperience}, {questionTypes}

Best Practice Tip: Keep surveys anonymous to encourage honest student feedback.

Enhanced Prompt:

Act as a 'Student Feedback Architect' and 'Instructional Improvement Coach'. Your task is to create a comprehensive student survey for [SUBJECT/GRADE LEVEL] to be administered [e.g., mid-year, end of a unit, end of semester]. The goal is to craft a short, powerful set of questions that provides authentic, actionable feedback on the learning experience, focusing on the student's perspective of their own understanding, the class-room environment, and the effectiveness of teaching practices.

The guide must be structured as follows:

1   **The "Why" it Matters**: A brief, impactful explanation of the core principle behind gathering student feedback (e.g., "Students are the primary consumers of our teaching; their perceptions are a critical data point for improving our practice. A well-designed survey anonymizes their voice, providing honest, crucial insights into what is working and what isn't. This transforms them from passive recipients into active partners in the learning process and provides us with a roadmap for our own professional growth.").

2   **Overarching Goal**: State the primary objective of this survey (e.g., "To gather confidential, structured feedback on student comfort, clarity, and engagement in order to identify specific, high-impact adjustments to my teaching methods, classroom routines, and learning environment for the remainder of the year.").

3   **The Survey**: A Balanced Mix for Quantitative and Qualitative Data Provide a list of 5-7 distinct questions. The questions must be strategically designed to be:

   ◇ **Quantitative (Scaled)**: For easy aggregation and identification of trends.

   ◇ **Qualitative (Open-Ended)**: For rich, nuanced detail and specific suggestions.

   ◇ **Focused on Key Levers**: Questions should target specific, adjustable elements of teaching (e.g., pace, clarity, support, classroom culture).

4   **Breakdown for Each Survey Question**: For every question, include:

◇ **The Question & Format**: Phrased clearly and neutrally. Specify the format (e.g., "Scale of 1-5 (1=Strongly Disagree, 5=Strongly Agree)", "Multiple Choice", "Short Answer").

◇ **The "Why" it's Important**: A brief explanation of what specific aspect of teaching or learning this question assesses and why that data is valuable (e.g., "This measures perceived psychological safety, which is the foundation for academic risk-taking and asking questions. A low score here is a critical flag that must be addressed before academic interventions can be fully effective.").

◇ **How to Analyze the Response**: A quick note on what to look for in the answers and what a positive or negative trend might indicate.

5 **The "Why" Behind This Strategy**: A concluding summary explaining how this short, focused approach to student feedback increases response rates, yields more honest and useful data than long, generic surveys, and empowers the teacher to make precise, evidence-based changes that directly respond to student needs.

6 **Bonus**: Pro-Tip for Teachers - Administering for Honest Feedback:

◇ **Anonymity is Non-Negotiable**: Stress that the survey is completely anonymous to ensure candid responses. Use a digital tool like Google Forms that doesn't collect emails.

◇ **Set the Stage**: Explain why you are giving the survey: "Your feedback helps me be a better teacher for you. I will read every response and use them to make our class better."

◇ **Close the Loop**: After reviewing the data, share back the overarching trends and one thing you will change based on their feedback. This shows you listened and validates their voice.

◇ **Act on the Data**: Pick 1-2 key pieces of feedback to act on immediately. This builds trust and makes the process meaningful.

The final output should be a complete, plug-and-play feedback kit that is easy to administer, simple to analyze, and packed with actionable insights.

Complete Output:

The "Why" it Matters: We often guess what our students are thinking. This survey replaces guesswork with data. By intentionally soliciting anonymous feedback, we demonstrate humility and a commitment to growth. The insights gathered are invaluable for differentiating instruction, building a stronger classroom community, and ensuring our teaching methods are effectively reaching every learner.

The Student Survey: Your Voice Matters

Instructions: This survey is 100% anonymous. Please be honest so I can understand what's working well and what I can improve to make our class a better place for learning. Thank you for your thoughtful feedback!

1 **Question**: On a scale of 1-5, how clearly do I explain new concepts and directions? (1=Very Unclear, 5=Very Clear)

   ▪ **Format**: Scale of 1-5

   ▪ **Why It's Important**: This directly assesscs the clarity of your direct instruction, which is the foundation for all subsequent learning activities. A low average score indicates a need to slow down, use more examples, or provide written instructions to complement verbal ones.

- **How to Analyze**: Look for the average score. Read any written comments for specifics on what was unclear.

2. **Question**: On a scale of 1-5, how comfortable do you feel asking me for help when you are confused? (1=Very Uncomfortable, 5=Very Comfortable)

- **Format**: Scale of 1-5

- **Why It's Important**: This measures the perceived "approachability" of the teacher and the psychological safety of the classroom. It is a prerequisite for effective differentiation and support.

- **How to Analyze**: A score below 4 is a critical red flag. This must be addressed by explicitly inviting questions, circulating more, and responding positively to all queries.

3. **Question**: Which activity type helps you learn the most? (Please choose one)

- A) Working quietly by myself

- B) Working with a partner or small group

- C) Whole-class discussions

- D) Hands-on projects or labs

- E) Watching videos or demonstrations

- **Format**: Multiple Choice

- **Why It's Important**: This provides crucial data on learning preferences, helping you balance your lesson planning to engage different types of learners. It's not about catering to preferences exclusively, but about ensuring variety.

- **How to Analyze**: Look for the most popular choice. Ensure your upcoming units include a mix of these modalities, especially the top ones.

4. **Question**: What is one thing I should START doing to help you learn better?

   - **Format**: Short Answer

   - **Why It's Important**: This open-ended question often yields the most specific and innovative suggestions for improvement. It frames feedback constructively and focuses on the future.

   - **How to Analyze**: Look for patterns. If multiple students suggest "more review games" or "check-ins," that is a clear action item.

5. **Question**: What is one thing I should STOP doing? It's not helping.

   - **Format**: Short Answer

   - **Why It's Important**: This is the hardest but most valuable question. It uncovers blind spots and unintended negative consequences of well-intentioned practices (e.g., "stop rushing through the slides," "stop cold-calling").

- **How to Analyze**: Read these with an open mind. Look for trends, but also be prepared for outlier comments that might still contain a kernel of truth.

6. **Question**: What is one thing you are proud of YOURSELF for learning or accomplishing in this class so far?

- **Format**: Short Answer

- **Why It's Important**: This ends the survey on a positive, metacognitive note. It reveals what students value and what they feel successful at, which is great data for building confidence and celebrating wins.

- **How to Analyze**: Use these responses to celebrate! Share anonymized successes with the class to build a culture of achievement.

The "Why" Behind This Strategy: This plan works because it respects everyone's time. The survey is short enough to complete in 5 minutes but deep enough to provide a goldmine of information. The mix of scaled and open-ended questions allows you to quickly spot trends (from the numbers) and then understand the "why" behind them (from the comments). This creates a clear, student-informed agenda for your own professional refinement.

Bonus: Pro-Tip for Teachers - Administering for Honest Feedback:

- **Do It Digitally**: Use Google Forms or Microsoft Forms. The platform automatically aggregates the quantitative data into charts for you, making analysis instantaneous.

- **Pick the Right Time**: Administer the survey during a neutral, calm time—not right before a break or after a test.

- **Model Non-Defensiveness**: When you share back the results, thank students for their honesty. If a criticism is hard to hear, acknowledge it without being defensive: "I saw that several of you find the pace too fast. I will work on that. Thank you for telling me." This models maturity and a growth mindset.

Prompt: Classroom Assessment Techniques

Simple Prompt: "Act as a 'Formative Assessment Coach' and 'Instructional Strategist'. Your task is to create a comprehensive guide to Classroom Assessment Techniques (CATs) for educators".

- **Tone**: Practical, Supportive
- **Format**: Technique List (x5)
- **Platform**: Classroom, Teacher Workshops
- **AI Role**: Formative Assessment Coach / Instructional Strategist
- **Prompt Goal**: Provide easy tools for real-time student feedback
- **Tags**: #assessmentTechniques, #formative, #teachingTools
- **Prompt Variables**: {topic}, {grade/level}, {learningGoals}

Best Practice Tip: Use peer-assessment occasionally to help students learn from one another.

Enhanced Prompt:

Act as a 'Formative Assessment Coach' and 'Instructional Strategist'. Your task is to create a comprehensive guide to Classroom Assessment Techniques (CATs) for educators. The goal is to provide a toolkit of low-effort, high-impact strategies that provide real-time evidence of student understanding, allowing teachers to adjust instruction during the learning process, not just at the end.

The guide must be structured as follows:

1   **The "Why" it Matters**: A brief, impactful explanation of the core principle behind formative assessment (e.g., "Formative assessment is the compass, not the autopsy. It is the ongoing process of gathering evidence of learning while teaching is still happening. This real-time feedback loop is the most powerful tool a teacher has to close the gap between what students currently understand and the desired learning goal, ensuring that no student is left behind due to misunderstood instruction.").

2   **Overarching Goal**: State the primary objective of using these techniques (e.g., "To move from teaching on autopilot to teaching with intention. These techniques provide a constant pulse on student understanding, enabling teachers to make informed, in-the-moment decisions about whether to re-teach, pivot, or move on, thereby maximizing the effectiveness of every instructional minute.").

3   **The CATs Toolkit**: A Strategic Mix Provide a list of 5-7 distinct techniques. The techniques must be categorized by their primary function:

   ◇   **Quick Checks for Understanding**: To gauge whole-class comprehension in seconds.

   ◇   **Promoting Metacognition**: To get students to think about their own thinking.

   ◇   **Encouraging Discussion & Dialogue**: To assess understanding through conversation.

4   **Breakdown for Each Assessment Technique**: For every technique, include:

   ◇   **Technique Name**: A clear, memorable name.

◇ **Category**: Label it as Quick Check, Metacognition, or Discussion.

◇ **How It Works**: A concise, step-by-step explanation of how to implement it in a live classroom setting.

◇ **The "Why" it's Important**: A brief explanation of the specific cognitive or pedagogical benefit (e.g., "This technique provides anonymous, whole-class data instantly, allowing the teacher to see if one student is confused or if the entire class missed the concept, without putting any individual student on the spot.").

◇ **What to Do with the Data**: The critical next step—how a teacher should respond to the information gathered.

5 **The "Why" Behind This Strategy**: A concluding summary explaining how this ongoing, low-stakes approach to assessment reduces student anxiety, creates a culture of mistake-making and growth, and transforms teaching from a one-size-fits-all lecture into a responsive, differentiated practice.

6 **Bonus**: Pro-Tip for Teachers - The Cycle of Inquiry:

◇ **The Loop**: Remember the cycle: Teach -> Assess (using a CAT) -> Analyze the data -> Adjust -> Re-teach (if needed). The CAT is useless without the action that follows.

◇ **Start Small**: Choose one technique to use once per day for a week. Master it before adding another.

◇ **No-Grade Zone**: The power of formative assessment is negated if it's graded. Keep it low-stakes to encourage honest responses.

◇ **Think Beyond Thumbs**: While thumbs up/down is a start, strive for techniques that provide more nuanced data (e.g., why a student is confused).

The final output should be a complete, plug-and-play assessment toolkit that empowers educators to become more responsive and data-informed in their teaching practice.

Complete Output:

The "Why" it Matters: Summative assessments (tests, quizzes) diagnose what students learned after the fact. Formative assessment, through Classroom Assessment Techniques, is the ongoing monitoring of the learning process. It's the difference between a doctor performing an annual physical (summative) and a doctor continuously monitoring a patient's vitals during surgery (formative). The latter allows for immediate intervention, which is precisely what effective teaching requires.

The CATs Toolkit: Your Pulse on Student Learning

1. Technique: The Exit Ticket

- **Category**: Quick Check for Understanding
- **How It Works**: In the last 3-5 minutes of class, students answer a single question or solve one problem on a small slip of paper. It must be turned in as they 'exit.' The prompt can be: 'What was the most important thing you learned today?' 'What question do you still have?' or a direct problem to solve.
- **Why It's Important**: It provides a perfect snapshot of understanding at the end of a lesson. It forces students to retrieve and synthesize information, and it gives the teacher a crucial planning tool for the next day's lesson.

- **What to Do with the Data**: Sort the tickets into three piles: "Got it," "Almost there," "Not there." Use this to form strategic groups for the next day's warm-up or to decide if the whole class needs a re-teach.

2. Technique: Four Corners

- **Category**: Quick Check for Understanding / Discussion

- **How It Works**: Pose a question with four possible answers (A, B, C, D) or four levels of agreement (Strongly Agree, Agree, Disagree, Strongly Disagree). Assign each answer a corner of the room. Students physically move to the corner that represents their answer.

- **Why It's Important**: It incorporates movement (kinesthetic learning) and instantly visualizes the distribution of understanding or opinion across the entire class. It's a great way to spark debate and discussion.

- **What to Do with the Data**: Ask students in different corners to defend their reasoning. This allows students to teach each other. If most of the class is in the wrong corner, it's a clear signal for immediate whole-class re-teaching.

3. Technique: Think-Pair-Share

- **Category**: Discussion / Metacognition

- **How It Works**: Pose a challenging, open-ended question. First, give students individual time to Think and jot down ideas. Then, have them Pair up with a partner to discuss their thoughts. Finally, Share their or their partner's ideas with the whole class.

- **Why It's Important**: It ensures all students have time to process and formulate an answer, not just the quickest hand-raisers. It builds language skills and allows for low-risk practice before sharing with the entire class.

- **What to Do with the Data**: Circulate during the "Pair"

phase to listen in on conversations. This is where you hear the real misconceptions and brilliant ideas. Use what you hear to select which pairs will share and to guide the whole-class discussion.

4. Technique: One-Minute Paper

- **Category**: Metacognition

- **How It Works**: Mid-lesson, pause and give students exactly one minute to answer a targeted prompt: "What is the clearest point so far?" "What is the muddiest point so far?" or "How does this connect to what we learned yesterday?"

- **Why It's Important**: It provides a "check-in" during the learning process, not just at the end. It teaches students to monitor their own understanding in real-time (metacognition) and gives you a chance to clarify confusion before moving on to the next concept.

- **What to Do with the Data**: Quickly scan responses. If you see a common "muddy point," stop and address it immediately. This prevents small confusions from snowballing into major misunderstandings.

5. Technique: Mini Whiteboards

- **Category**: Quick Check for Understanding

- **How It Works**: Every student has a small whiteboard, marker, and eraser. Pose a question. All students write their answer and hold it up on your count.

- **Why It's Important**: It guarantees 100% participation and provides immediate, visible data from every single student. It's perfect for quick math problems, vocabulary, short answers, or multiple-choice questions.

- **What to Do with the Data**: Do a quick visual scan. If you see a wide variety of answers, have students turn and convince their neighbor they are right, then revote.

This promotes academic discourse and self-correction.

The "Why" Behind This Strategy: This plan works because it treats assessment as a teaching tool, not a grading tool. These techniques are fast, low-stakes, and embedded directly into instruction. They shift the classroom culture from "Did you get the right answer?" to "How are we all thinking about this?" This creates a safe environment for risk-taking and provides the teacher with the actionable data needed to be truly responsive to student needs.

Bonus: Pro-Tip for Teachers - The Cycle of Inquiry:

- **The Most Important Step**: The assessment itself is pointless without the instructional adjustment. Always ask yourself: "Based on what I just learned, what will I do next?" This is the core of responsive teaching.

- **Variety is Key**: Don't use the same technique every day. Rotate them to keep students engaged and to assess different types of thinking.

- **Involve Students**: Sometimes, after using a technique, share the aggregated data with the class. Say, "I noticed about half of us are confident with this, and half is still unsure. Let's pair up so you can teach each other." This builds a collaborative learning community.

Prompt: Peer Feedback Activity Designer

Simple Prompt: "Act as a 'collaboration mentor'. Your role involves designing a peer feedback activity for [assignment/project] that enables students to engage thoughtfully with one another's work and to cultivate effective strategies for delivering meaningful, supportive critiques.

- **Tone**: Fair, Encouraging
- **Format**: Rubric + Feedback Section
- **Platform**: Assignments, LMS, Report Cards
- **AI Role**: Assessment for Learning Coach / Transparent Design Specialist
- **Prompt Goal**: Support fair grading while leaving room for individual feedback
- **Tags**: #grading, #feedback, #rubric
- **Prompt Variables**: {assignment/project}, {criteria}, {feedbackStyle}

Best Practice Tip: Leave at least one open-ended "teacher note" section for student-specific encouragement.

Enhanced Prompt:

You are an 'Educational Ethologist' and 'Classroom Culture Architect'. Your expertise lies in designing structured interactions that cultivate academic discourse, metacognition, and a growth mindset. You understand that peer feedback is not an add-on activity but a core pedagogical strategy for deepening learning, developing critical thinking, and fostering a supportive community of learners.

Primary Objective:

To design a complete, ready-to-implement protocol for a peer feedback activity that transforms the classroom into a community of practice, where students learn to give and receive constructive, specific, and kind feedback as a natural part of the learning process.

Activity Parameters:

1. **Subject & Assignment**: [e.g., "8th Grade Science Lab Report," "10th Grade Literary Analysis Essay," "Group Project Presentation on Ancient Civilizations"]

2. **Learning Objective for the Feedback Giver**: [e.g., "Identify and articulate evidence-based claims," "Evaluate the clarity of a scientific procedure," "Assess the strength of supporting evidence"]

3. **Learning Objective for the Feedback Receiver**: [e.g., "Identify patterns in feedback to guide revision," "Practice active listening and note-taking," "Develop resilience and a growth mindset"]

4. **Grade Level**: [e.g., Upper Elementary, Middle School, High School]

Core Requirements for the Feedback Protocol:

The protocol must be a multi-phase, scaffolded system that includes the following components:

1. The Pedagogical Foundation ("The Why"):
- **Articulate the core principle**: "Peer review is where learning becomes visible." When students evaluate another's work against a clear standard, they develop a sharper critical eye for their own. This process external-

izes the internal checklist of quality, accelerating mastery for both the giver and receiver of feedback.

2. The Calibrated Rubric (The "What"):

- Provide a simple, student-friendly rubric with 3-4 criteria directly tied to the assignment's core learning objectives.

- **Format**: Use a single-point rubric or a 3-level rubric (e.g., "Developing," "Proficient," "Exemplary") for simplicity.

- **Focus on Observable Traits**: Criteria should be specific and observable, not vague (e.g., instead of "Good conclusion," use "Conclusion restates the thesis and summarizes the main evidence").

- Include space for "Glow" (What worked well) and "Grow" (A specific suggestion for improvement) comments for each criterion.

3. The Structured Activity Architecture (The "How"):

- Design a timed, step-by-step workshop structure to ensure productivity and focus.

   ◇ **Phase 1**: Setup & Norming (10 mins): Teacher introduces the rubric and models feedback with a sample anonymous work product ("I Do").

   ◇ **Phase 2**: Calibration (5 mins): In pairs, students practice applying the rubric to the same sample product to calibrate their understanding ("We Do").

   ◇ **Phase 3**: Focused Feedback Exchange (15-20 mins): Students swap work and use the rubric to provide written feedback. Use a timer to keep them on task.

   ◇ **Phase 4**: Debrief & Reflection (10 mins): Receivers share one piece of feedback they found most useful. Givers share what they learned about quality by evaluating someone else's work.

4. The Sentence Stems & Communication Guide (The "Language"):

- Provide a "Feedback Menu" of categorized sentence stems to scaffold respectful and actionable language.

  ◇ **"Glow" Stems (To Praise)**: "I really liked how you..." / "Your argument was strong when you..." / "The way you [specific example] was effective because..."

  ◇ **"Grow" Stems (To Improve)**: "I suggest... because..." / "Consider... to make this even clearer." / "A question I had was... could you clarify?"

  ◇ **Prohibition**: Explicitly discourage vague or personal language (e.g., "This is bad," "I don't like it").

5. The Facilitation Guide for the Teacher (The "Management"):

- **Provide a simple 3-step guide for the teacher to ensure success:**

  ◇ **Frame the Mindset**: Begin by stating, "We are a community of learners helping each other grow. Our goal is to be kind, specific, and helpful."

  ◇ **Monitor & Coach**: Circulate during the activity, listening for the use of target language and intervening to re-direct unproductive feedback.

  ◇ **Close the Loop**: Require students to submit their reviewed rubric with their final draft, explaining how they used the peer feedback in their revisions. This creates accountability.

6. The Metacognitive Rationale ("The Why Behind the Strategy"):

- **Explain the deep learning benefits for students**:

  ◇ **For the Giver**: Evaluating another's work is a high-level cognitive task (Bloom's Evaluation) that reinforces their own understanding of quality and criteria.

  ◇ **For the Receiver**: It provides multiple perspectives on their work, breaking them out of their own cognitive fixedness and providing a roadmap for revision.

  ◇ **For the Class**: It builds a culture of collective intelligence and shared ownership over learning, reducing reliance on the teacher as the sole source of judgment.

Final Output Structure: I require the final protocol to be presented as a ready-to-use teacher's guide.

Output Format:

- **Title**: The Collaborative Feedback Protocol: [Assignment Name]

- **Header**: Grade Level, Time Needed, Materials

- **Section 1**: Teacher Script (A brief script for introducing the activity and its purpose)

- **Section 2**: The Calibrated Rubric (A printable rubric template)

- **Section 3**: Activity Timer (A step-by-step breakdown with suggested times)

- **Section 4**: Student Feedback Menu (The list of sentence stems)

- **Section 5**: Differentiation Tips: How to support struggling students and extend advanced students during the process.

- **Pro-Tip**: "Always have students review work anonymously if possible to reduce bias. If not, carefully consider pairings to ensure a productive and safe environment."

Final Request: Please begin by asking me for the Subject & Assignment, Learning Objectives, and Grade Level. Then, generate the complete Collaborative Feedback Protocol.

Prompt: Letter of Recommendation

Simple Prompt: "Act as an 'Student Advocacy Specialist' and 'Academic Mentor'. Your task is to create a comprehensive guide and template for writing a compelling letter of recommendation for a [STUDENT NAME] applying for [OPPORTUNITY: e.g., college, scholarship, internship, job]".

- **Tone**: Professional, Supportive
- **Format**: Recommendation Letter
- **Platform**: University Applications, Scholarships
- **AI Role**: Student Advocacy Specialist / Academic Mentor
- **Prompt Goal**: Provide personalized and compelling student recommendations.
- **Tags**: #recommendation, #studentSupport, #applications
- **Prompt Variables**: {studentName}, {program/scholarship}, {achievements}

Best Practice Tip: Keep the letter personal by mentioning specific examples of the student's work or character.

Enhanced Prompt:

Act as a 'Student Advocacy Specialist' and 'Academic Mentor'. Your task is to create a comprehensive guide and template for writing a compelling letter of recommendation for a [STUDENT NAME] applying for [OPPORTUNITY: e.g., college, scholarship, internship, job]. The goal is to produce a letter that moves beyond generic praise to provide a vivid, evidence-based portrait of the student's unique character, intellectual curiosity, and potential for future success, ultimately persuading the selection committee to choose them.

The guide must be structured as follows:

1   **The "Why" it Matters**: A brief, impactful explanation of the core principle behind a powerful recommendation (e.g., "A standout letter of recommendation is a narrative, not a list. It provides a trusted, third-party validation of a student's application by telling a specific story of their growth, character, and impact. It answers the committee's most critical question: 'Beyond grades and test scores, what makes this person truly exceptional and a perfect fit for our community?'").

2   **Overarching Goal**: State the primary objective of this letter (e.g., "To provide a compelling, anecdote-driven argument for [Student Name]'s admission/selection by highlighting their unique intellectual presence, resilience, and capacity for leadership and contribution.").

3   **The Letter Template**: A Persuasive Structure The template must follow a formal business letter format and include these key sections:

◇   **Header & Salutation**: Your contact info, date, and a formal address.

◇   **Opening Paragraph**: The "Hook" Your qualifications and context, followed by a powerful, overarching statement of endorsement.

◇   **Body Paragraph 1**: The Academic Portrait A specific story that illustrates their intellectual character.

◇   **Body Paragraph 2**: The Personal Character Portrait A specific story that illustrates their soft skills and impact on others.

◇ **Body Paragraph 3**: The "Fit" Paragraph A direct connection between the student's qualities and the specific values/needs of the opportunity.

◇ **Closing Paragraph**: The Unqualified Recommendation A strong, comparative summary and your contact offer.

◇ **Signature**: Formal closing.

4 **The "How-To"**: A Simple 3-Step Guide to Personalization

◇ **Step 1**: Gather Evidence. Before writing, ask the student for their resume, personal statement, and a bulleted list of their top 2-3 accomplishments in your class/activity. Ask them, "What is the one thing you most want me to highlight?"

◇ **Step 2**: Find the Anecdote. For each key trait (e.g., resilience, curiosity), identify one specific moment, project, or interaction that proves it. Show, don't just tell.

◇ **Step 3**: Align with the Opportunity. Research the program/company. Weave in keywords from their mission statement or course descriptions to demonstrate a perfect fit.

5 **The Call to Action (CTA)**: A strong, clear, and direct call-to-action.

◇ **Example CTA**: "[Student Name] is not just a promising student; they are a future leader. I recommend them to you without any reservation and am confident they will exceed your highest expectations. I welcome the opportunity to discuss their qualifications with you further. Please do not hesitate to contact me at [Your Phone Number] or [Your Email]."

6 **The "Why" Behind This Strategy**: A concluding summary explaining how this structured, story-based approach transforms a standard form letter into a decisive asset for the student's application. It provides the committee with the concrete evidence and vivid narrative they need to advocate for the student during deliberations.

7 **Bonus**: Pro-Tip for Teachers - The Gold Standard:

   ◇ **The "Top 1%" Statement**: Compare the student to their peers. Phrases like "In my 15 years of teaching, [Student Name] ranks among the top 5%..." are incredibly powerful.

   ◇ **Avoid Clichés**: Strike words like "hard-working" and "nice" from your vocabulary. Replace them with "perseverant," "intellectually tenacious," "collaborative," or "ethically grounded."

   ◇ **Quantify When Possible**: "They improved their grade from a B to an A through relentless revision" is better than "they worked hard."

The final output should be a complete, plug-and-play recommendation kit that empowers you to write a letter that truly makes a student's application stand out.

Complete Output:

The "Why" it Matters: A letter of recommendation is a character witness statement. Its power lies in its specificity and the credibility of the writer. A committee reads hundreds of applications; a letter that tells a memorable story about a student's character, curiosity, and resilience provides the human context that test scores and grades cannot, often becoming the deciding factor.

The Letter Template: A Persuasive Structure

[Your Name]

[Your Title, e.g., AP Biology Teacher]

[Your School Name]

[School Address]

[Your Email] | [Your Phone Number]

[Date]

[Admissions Committee / Hiring Manager Name]

[Title]

[Organization Name]

[Organization Address]

Dear [Mr./Ms./Mx. Last Name] / [Admissions Committee],

It is with immense pleasure and my highest recommendation that I write to you in support of [Student Name]'s application to [Name of Program/University/Job]. In my [Number] years as a [Your Subject] teacher at [Your School], I have encountered few students with [Student Name]'s combination of intellectual vitality, genuine curiosity, and unwavering integrity.

Paragraph 1: The Academic Portrait

My first vivid memory of [Student Name] was during our unit on

[Specific Topic]. While other students were content to memorize the formulas, [Student Name] stayed after class to ask how these principles applied to [A Complex Real-World Problem]. This wasn't for extra credit; it was a genuine thirst for understanding. This intellectual tenacity is a hallmark of their work. For their final project, they [Describe a specific, impressive project—e.g., "designed and ran a novel experiment on..."], demonstrating not just mastery of content, but a remarkable capacity for independent, critical inquiry that is rare in a student of their age.

Paragraph 2: The Personal Character Portrait

Beyond their academic prowess, [Student Name] elevates everyone around them. I recall a moment when a group project was faltering due to conflicting ideas. [Student Name] did not dominate; instead, they facilitated a conversation, synthesizing disparate viewpoints and helping the group find a consensus that incorporated the best of everyone's thinking. They are a natural collaborator—respectful, empathetic, and driven by a desire to see the whole team succeed. Their positive attitude and innate leadership are intangible assets that will greatly enrich your community.

Paragraph 3: The "Fit" Paragraph

[Student Name]'s specific passion for [Student's Interest] aligns perfectly with [Name of Program/University]'s renowned focus on [Program's Specialty or a specific course]. The way they have already begun to [Something the student has done related to the program] demonstrates a clear trajectory that your program is designed to nurture. I have no doubt they will not only thrive in your rigorous academic environment but will also contribute meaningfully to campus discussions and collaborations.

Closing Paragraph: The Unqualified Recommendation

In a cohort of high-achieving students, [Student Name] stands out as truly exceptional. They possess the maturity, intellectual creativity, and personal character to make a significant impact at your institution and beyond. I recommend them to you without any reservation.

Call to Action (CTA):

I am confident that [Student Name] will be a transformative member of your community. I welcome the opportunity to provide further context; please feel free to contact me at [Your Phone Number] or [Your Email].

Sincerely,

[Your Signature (for a hard copy)]

[Your Typed Name]

[Your Title]

The "How-To": A Simple 3-Step Guide to Personalization

- **Replace the Brackets**: Fill in all [ ] placeholders with specific details.

- **Insert Your Anecdotes**: Replace the example stories in Paragraphs 1 and 2 with your own specific, vivid memories of the student. This is the most critical step.

- **Customize the "Fit"**: Research the specific program/job and tailor Paragraph 3 to show why this student is a perfect match for that specific opportunity.

The "Why" Behind This Strategy: This plan works because it provides a structure for what selection committees value most: specific, story-based evidence of character and intellect. It forces the writer to move beyond bland adjectives and instead provide a narrative that proves the student's qualities, making the letter memorable, credible, and powerfully persuasive.

Bonus: Pro-Tip for Teachers - The Gold Standard:

- **The "Fizz" Test**: Ask yourself: "Does this letter have 'fizz'?" Does it sparkle with specific details that make the student come alive? If it feels generic, add more story.

- **Read It Aloud**: The rhythm and flow of the letter matter. Reading it aloud will help you catch awkward phrasing and ensure it sounds confident and compelling.

- **Let the Student See a Draft**: This is considered best practice. It ensures accuracy and allows them to jog your memory about other accomplishments you can include.

Prompt: Resume Enhancer (Student/Academic Version)

Simple Prompt: "Act as a You are an expert 'Career Strategy Consultant' and 'Academic Resume Specialist'. Your task is to perform a strategic rewrite and enhancement of a student's resume, transforming it from a generic list of experiences into a targeted, achievement-focused document tailored to a specific, high-value opportunity".

- **Tone**: Professional, Polished
- **Format**: Resume Rewrite
- **Platform**: Student Applications, Job/Internship Portfolios
- **AI Role**: Career Strategy Consultant / Academic Resume Specialist
- **Prompt Goal**: Optimize student resumes for admissions or career opportunities
- **Tags**: #resume, #studentSuccess, #careerSupport
- **Prompt Variables**: {program/scholarship/role}, {studentBackground}, {skills}

Best Practice Tip: Encourage students to track achievements throughout the year so resumes stay updated.

Enhanced Prompt:

You are an expert 'Career Strategy Consultant' and 'Academic Resume Specialist' with a focus student and early-career candidates. Your expertise lies in reframing academic and extracurricular experiences into quantifiable, impact-driven narratives that resonate with specific opportunity reviewers and bypass automated screening systems.

Primary Objective: To perform a strategic rewrite and enhancement of a student's resume, transforming it from a generic list of experiences into a targeted, achievement-focused document tailored to a specific, high-value opportunity.

Input Parameters:

- **Target Opportunity**: [e.g., "Summer Data Science Internship at Netflix," "The Fulbright Scholarship," "Undergraduate Research Program in Neuroscience," "Marketing Coordinator Role at a Startup"]

- **Student's Current Resume**: [I will paste the current resume text here]

- **Opportunity Description/Key Requirements**: [I will paste the job description, scholarship criteria, or program mission statement here]

Core Requirements for the Resume Transformation: The transformation must be a comprehensive overhaul, not just minor edits. The final output must include the following components:

1. Strategic Narrative Development:

- **Customized Professional Summary**: Craft a powerful 2-3 line summary that functions as an elevator pitch.

- **Must Include**: The target role/program, 2-3 key strengths or areas of expertise (directly mirroring the opportunity's requirements), and a statement of objective or value proposition.

- **Tone**: Confident, professional, and specific.

2. Experience Reframing & Impact Quantification:

- Identify the 5-7 most relevant experiences from the student's current resume (including coursework, projects, clubs, volunteer work, and part-time jobs).

- For each selected experience, rewrite 1-2 bullet points to be achievement-oriented.

    ◇ **Formula**: Strong Action Verb + What You Did + Quantifiable Result or Impact.

    ◇ **Example Transformation**:

        · **Before**: "Responsible for managing social media accounts."

        · **After**: "Developed and executed a content calendar, increasing follower engagement by 25% over a 3-month period."

3. ATS & Keyword Optimization Audit:

- Analyze the provided opportunity description.
- Extract 10-15 critical hard and soft skills keywords (e.g., "Python," "data visualization," "cross-functional collaboration," "quantitative analysis").
- Provide a list of these keywords and explicitly state where each one has been integrated into the revised resume (e.g., "Keyword 'SPSS' integrated into Skills section and Project A bullet point").

4. Skills Section Architecture:

- Reorganize the skills section into clearly defined, scannable categories.
- **Recommended Categories**: Technical Skills, Laboratory Skills, Software & Programming, Languages, Relevant Coursework, Certifications, Soft Skills.
- For key skills, add brief context where possible (e.g., "Python (Pandas, NumPy)", "Public Speaking (led presentations for groups of 50+)").

5. Rationale Dossier:

- This is critical. For each major change (Summary, key bullet points, skills organization), provide a brief 'Strategic Reasoning' note.

- **Explain**: Why this change was made and how it strengthens the candidate's alignment with the specific target opportunity. (e.g., "Changed 'cashier' to 'Customer Service Representative' to better reflect the developed soft skills relevant to a client-facing role.").

Final Output Structure: I require the final output to be presented in two clear parts.

Part A: The Enhanced Resume

- A complete, formatted, and polished resume document, ready to be copied and pasted.

- **It should include all standard sections**: Contact Info, Professional Summary, Education, Experience, Projects, Skills, and Relevant Extracurriculars.

Part B: The Consultant's Rationale Report

- **A separate section that details the strategic choices made, including**:

  - ◇ **Keyword Audit Results**: The list of extracted keywords and their placement.

  - ◇ **Change Log**: A bulleted list of the key transformations with their corresponding strategic reasoning.

  - ◇ **Pro-Tips for the Student**: 2-3 final pieces of advice (e.g., "Be prepared to discuss Project X in detail during an interview," "Consider adding a link to your GitHub portfolio next to your email.").

Final Request:

Please begin by asking me for the Target Opportunity, the Student's Current Resume, and the Opportunity Description. Then, generate the complete Strategic Student Resume Transformation as outlined.

Category 4: Teacher Productivity & Support

Theme:

This category is dedicated to providing teachers with the tools and strategies necessary to manage their time, classrooms, and professional growth effectively. By integrating proven classroom management techniques with personal productivity tools, the goal is to help educators achieve an optimal balance between their workload and personal well-being.

Through these combined approaches, teachers are better equipped to reduce stress and enhance their overall impact in the classroom. The focus is on supporting educators as they strive to maximize their effectiveness while maintaining a sustainable and fulfilling career in education.

Category Goal:

Educators benefit greatly from the implementation of practical productivity systems and support frameworks within the classroom environment. These systems are designed to streamline classroom management processes, allowing teachers to efficiently organize tasks and maintain order. By utilizing these frameworks, teachers can effectively save time on daily responsibilities, enabling them to focus more on instruction and student engagement. Additionally, the integration of continuous professional development opportunities encourages educators to refine their skills and remain current with best practices in education.

Overall, these strategies contribute to a more organized, productive, and supportive learning environment, empowering teachers to maximize their impact while maintaining personal well-being.

Mini-Index: Teacher Productivity & Support

Prompt	AI Role	Prompt Goal
Classroom Management Strategies	Assessment Designer	Provide practical strategies for behavior management
Classroom Management Tools	Edtech Advisor	Recommend tools/apps for classroom efficiency
Classroom Management App Generator	Innovation Coach	Inspire ideas for classroom management apps
Behavior Management Techniques	Behavioral Coach	Offer evidence-based methods for handling disruptions
Time Management for Teachers	Productivity Consultant	Create balanced weekly schedules for teachers
Professional Development Goals	Teacher Coach	Suggest goals and growth plans for teachers
Classroom Organization Tips	Classroom Designer	Provide organization tips for smooth teaching
Classroom Icebreaker Questions	Engagement Coach	Generate icebreaker questions to build rapport
Daily Growth Blueprint (Teacher)	Personal Coach	Help teachers balance teaching, grading, and self-care
Educational Website Generator	Edtech Strategist	Propose website ideas to support teaching

Detailed Prompts

Prompt: Classroom Management Strategies

Simple Prompt: "Act as an 'Assessment Designer' and 'Curriculum Alignment Specialist'. Your task is to create a comprehensive guide to classroom management strategies for a [GRADE LEVEL] classroom".

- **Tone**: Supportive, Practical
- **Format**: Strategy List (x5)
- **Platform**: Teacher Guides, Training Materials
- **AI Role**: 'Assessment Designer/Curriculum Alignment Specialist
- **Prompt Goal**: Provide teachers with practical classroom management tools
- **Tags**: #classroomManagement, #teacherSupport, #behavior
- **Prompt Variables**: {grade/level}, {classSize}, {behavioralChallenges}

Best Practice Tip: Combine preventive measures (rules, routines) with positive reinforcement for better results.

Enhanced Prompt:

Act as an 'Assessment Designer' and 'Curriculum Alignment Specialist'. Your task is to create a comprehensive guide to classroom management strategies for a [GRADE LEVEL] classroom. The goal is to provide a toolkit of strategies that are proactive, relationship-based, and designed to create a safe, predictable, and productive learning environment where all students can thrive.

The guide must be structured as follows:

1 **The "Why" it Matters**: A brief, impactful explanation of the core philosophy behind effective management (e.g., "Classroom management is not about control; it is about the intentional design of a learning environment. The most effective strategies are proactive and preventative, rooted in strong teacher-student relationships, clear expectations, and student dignity. This approach minimizes the need for reactive discipline and maximizes instructional time.").

2 **Overarching Goal**: State the primary objective of this management philosophy (e.g., "To cultivate a classroom community built on mutual respect, shared responsibility, and a focus on learning, thereby creating the conditions for high academic and social-emotional growth.").

3 **The Strategic Toolkit**: A Tiered Approach Provide a list of 5-7 distinct strategies. The strategies must be categorized into a logical, tiered framework:

⋄ **Tier 1**: Proactive & Preventative (Foundation): Strategies to establish systems and prevent misbehavior before it starts.

⋄ **Tier 2**: Responsive & Supportive (Intervention): Strategies to address low-level misbehavior calmly and consistently.

⋄ **Tier 3**: Relationship-Restorative (Connection): Strategies to repair harm, rebuild trust, and understand the root cause of recurring behaviors.

4 **Breakdown for Each Strategy**: For every strategy, include:

◇ **Strategy Name**: A clear, actionable name.

◇ **Tier**: Label it as Proactive, Responsive, or Relationship-Restorative.

◇ **How It Works**: A concise, step-by-step explanation of how to implement the strategy in a classroom setting.

◇ **The "Why" it's Important**: A brief explanation of the psychological or pedagogical principle it leverages and its long-term impact on classroom culture (e.g., "This strategy uses non-verbal cues to correct behavior without disrupting the flow of a lesson, preserving student dignity and maintaining instructional momentum.").

5 **The "Why" Behind This Strategy**: A concluding summary explaining how this tiered, proactive approach is more sustainable and effective than a purely reactive model. It reduces teacher stress, builds student self-regulation, and creates a positive classroom climate where learning is the central focus.

6 **Bonus**: Pro-Tip for Teachers - The Key to Consistency:

◇ **Start Small**: "Choose one or two strategies to master first. Implement them with relentless consistency before adding more."

◇ **Teacher Mindset**: "Your calm is contagious. Regulate your own emotions first; a frustrated tone can escalate a situation, while a calm tone can de-escalate it."

◇ **The Power of "I Notice"**: "Use language that describes the desired behavior rather than highlighting the misbehavior (e.g., 'I notice group three is ready with their materials' instead of

'Why isn't everyone ready?')."

◇ **Plan for Re-teaching**: "Expect to re-teach procedures and expectations after weekends, holidays, and anytime the classroom energy shifts. This is not a sign of failure, but of good teaching."

The final output should be a complete, plug-and-play management toolkit that empowers educators to build a respectful and highly functional classroom community from day one.

Prompt: Classroom Management Tools

Simple Prompt: "Act as an expert in Pedagogical Design and Proactive Classroom Culture. Your task is to create a comprehensive guide to classroom management tools—tangible systems and procedures—for a [GRADE LEVEL] classroom".

- **Tone**: Informative, Practical
- **Format**: Tool Recommendation List
- **Platform**: LMS, Teacher Resources
- **AI Role**: Edtech Advisor
- **Prompt Goal**: Suggest technology that supports discipline, organization, and efficiency
- **Tags**: #classroomTools, #teacherSupport, #edtech
- **Prompt Variables**: {grade/subject}, {teacherNeeds}, {budget}

Best Practice Tip: Prioritize apps that are easy to set up and integrate with existing classroom systems.

Enhanced Prompt:

Act as an expert in Pedagogical Design and Proactive Classroom Culture. Your task is to create a comprehensive guide to classroom management tools—tangible systems and procedures—for a [GRADE LEVEL] classroom. The goal is to provide a toolkit of concrete, ready-to-implement tools that prevent misbehavior, promote self-regulation, and create a predictable, respectful, and efficient learning environment for all students.

The guide must be structured as follows:

1 **The "Why" it Matters**: A brief, impactful explanation of the core principle behind systematic management (e.g., "Classroom management tools are the operating system of your classroom. They are the pre-designed structures and routines that automate expectations, minimize ambiguity, and conserve your—and your students'—cognitive energy for learning, not for figuring out what to do next. The right tool transforms potential conflict into a simple, predictable process.").

2 **Overarching Goal**: State the primary objective of using these tools (e.g., "To establish clear, reliable management systems that automate routines, reduce misbehavior, and streamline classroom logistics, allowing teachers to spend less energy on discipline and more time on instruction and relationship-building").

3 **The Management Toolkit**: A Tiered Approach Provide a list of 5-7 distinct tools. The tools must be categorized by their primary function:

   ◇ **Proactive & Preventative Tools**: Systems to establish routines and prevent issues before they start.

   ◇ **Responsive & Redirective Tools**: Tools to address off-task behavior calmly and consistently.

   ◇ **Motivational & Reinforcement Tools**: Systems to recognize and celebrate desired behaviors.

4 **Breakdown for Each Management Tool**: For every tool, include:

   ◇ **Tool Name**: A clear, actionable name.

   ◇ **Category**: Label it as Proactive, Responsive,

or Motivational.

◇ **How It Works**: A concise, step-by-step explanation of how to implement the tool, including what to say and what physical materials are needed.

◇ **The "Why" it's Important**: A brief explanation of the specific problem it solves and the long-term executive functioning or social-emotional skill it builds (e.g., "This tool provides a non-verbal, non-disruptive cue for redirection, preserving student dignity and keeping the lesson's momentum flowing.").

5 **The "Why" Behind This Strategy**: A concluding summary explaining how this toolkit approach—feeling a toolbox with specific instruments rather than a single hammer—empowers the teacher to respond to any situation with calm confidence. It creates a predictable environment that reduces anxiety for students and stress for the teacher.

6 **Bonus**: Pro-Tip for Teachers - The Implementation Framework:

◇ **Teach the Tool**: Never assume students know how to use a system. Explicitly teach, model, and practice each tool as you would an academic skill.

◇ **Consistency is Key**: A tool used inconsistently is worse than no tool at all. Pick one or two to master and implement with fidelity before adding more.

◇ **Involve Students**: Have students help design the tools or choose rewards. Ownership increases buy-in exponentially.

◇ **Re-teach After Breaks**: Assume procedures need to be re-taught after weekends, holidays, or any break in routine. This is not a failure; it's good teaching.

The final output should be a complete, plug-and-play management toolkit that provides the concrete "how-to" for creating a classroom that runs smoothly and respectfully.

Complete Output:

The "Why" it Matters: The most effective classroom management is invisible, baked into the environment through predictable systems. These tools are not about control; they are about clarity. They answer the constant student questions of "What do I do?" "What do I need?" and "What happens if...?" before they are even asked, freeing up mental space for critical thinking and engagement.

The Management Toolkit: Your Go-To Systems

1. Tool: The "What To Do" Visual Agenda

- **Category**: Proactive & Preventative
- **How It Works**: Every day, post a clear agenda (on the whiteboard or a digital slide) that lists the day's activities in order, including timings. Use icons or images for younger students. Review it at the start of the day and reference it during transitions.
- **Why It's Important**: This reduces anxiety about the unknown and minimizes off-task behavior during transitions. It answers the "What are we doing next?" question permanently, building student confidence and executive functioning skills like time management.

2. Tool: Non-Verbal Signal System

- **Category**: Responsive & Redirective
- **How It Works**: Establish and teach hand signals for common needs. Examples:

  ◇ **Restroom**: Hand raised with fingers crossed.

  ◇ **Water**: Hand raised with a "W" sign.

  ◇ **I Need Help**: A small tent card or red cup on the desk.

  ◇ **"I'm Listening"**: Teacher raises hand; students stop talking and raise theirs in response.

- **Why It's Important**: This eliminates countless minor disruptions. Students can have their needs met without interrupting instruction or drawing attention, preserving the flow of the lesson and student dignity.

3. Tool: Strategic Timer

- **Category**: Proactive & Preventative
- **How It Works**: Use a large, visible timer (online or physical) for every activity with a time limit. Give time warnings ("5 minutes left," "1 minute left").
- **Why It's Important**: This creates a sense of urgency and purpose, dramatically improving on-task behavior and time management. It externalizes the timekeeper role, removing the teacher from the position of constantly nagging students to hurry up.

4. Tool: "Stop & Go" Light or Noise Meter

- **Category**: Responsive & Redirective / Proactive
- **How It Works**: Use a physical traffic light or a digital app (like ClassDojo's noise meter) to provide visual feedback on acceptable noise levels during different activities (e.g., red for silent work, yellow for partner whisper, green for group discussion).

- **Why It's Important**: This provides clear, objective, and non-confrontational feedback to the entire class about their volume. It helps students self-monitor and regulate their behavior, making them active participants in managing the classroom environment.

5. Tool: Classroom Jobs Chart

- **Category**: Proactive & Motivational

- **How It Works**: Assign specific, rotating jobs to every student (e.g., Materials Manager, Technology Chief, Librarian, Greeter). Use a chart with clothespins or magnets to make rotations easy.

- **Why It's Important**: This builds community, responsibility, and ownership. It saves the teacher countless small tasks and gives every student a valued role, increasing their investment in the classroom and their sense of belonging.

6. Tool: Positive Behavior "Spotlight" Frame

- **Category**: Motivational & Reinforcement

- **How It Works**: Keep a special picture frame on your desk. When you "catch" a student demonstrating a target behavior, take their photo and put it in the frame with a caption explaining what they did (e.g., "Maria for persevering on her math work!").

- **Why It's Important**: This provides visual, public recognition for positive behavior that isn't tied to material rewards. It celebrates character and effort, reinforces expectations for others, and creates a culture of positivity.

The "Why" Behind This Strategy: This plan works because it provides specific, replicable systems instead of abstract advice. Each tool addresses a common classroom pain point with a clear procedure. By implementing these tools, you shift from reacting to misbehavior to proactively designing a environment where it is less likely to occur. This is the difference between managing behavior and engineering a culture of learning.

Bonus: Pro-Tip for Teachers - The Implementation Framework:

- **Introduce One at a Time**: Roll out one new tool per week. Teach it, practice it, and perfect it before introducing another.

- **Co-Create Expectations**: For tools like the Noise Meter, have the class help define what "Level 1 (Silent)" and "Level 2 (Whisper)" sound like. They will be more likely to adhere to rules they helped create.

- **Your Calm is Contagious**: The most important management tool is your own regulated nervous system. Use these tools to help you stay calm and consistent.

Prompt: Classroom Management App Generator

Simple Prompt: "Act as an expert in Educational Technology and Pedagogical Systems. Your task is to create a curated list of classroom management apps for a [GRADE LEVEL] classroom".

- **Tone**: Innovative, Practical
- **Format**: App Concept Plan
- **Platform**: Edtech Development, Teacher Tools
- **AI Role**: Innovation Coach
- **Prompt Goal**: Inspire app ideas that solve classroom management challenges
- **Tags**: #appIdeas, #classroomManagement, #edtech
- **Prompt Variables**: {grade/level}, {featureFocus}, {integrationNeeds}

Best Practice Tip: Test the idea with teacher feedback before building to ensure adoption.

Enhanced Prompt:

Act as an expert in Educational Technology and Pedagogical Systems. Your task is to create a curated list of classroom management apps for a [GRADE LEVEL] classroom. The goal is to provide a toolkit of digital applications that streamline administrative tasks, promote positive behavior, facilitate clear communication, and provide actionable data—freeing the teacher to focus more on instruction and relationship-building.

The guide must be structured as follows:

1   **The "Why" it Matters**: A brief, impactful explanation of the core principle behind using management apps (e.g., "Classroom management apps are not about replacing teacher authority; they are about systematizing it. They automate repetitive tasks like attendance, grading, and behavior tracking, provide transparent and immediate feedback loops for students, and offer data-driven insights that allow teachers to intervene proactively rather than reactively. This transforms management from a drain on energy to a streamlined process.").

2   **Overarching Goal**: State the primary objective of integrating these digital tools (e.g., "To create a more efficient, transparent, and positive classroom ecosystem by leveraging technology to handle logistics, reinforce expectations, and strengthen the home-school connection, ultimately maximizing instructional time and student engagement.").

3   **The App Menu**: A Strategic Mix by Function Provide a list of 5-7 distinct apps. The apps must be categorized by their primary management function:

   ◇   **Behavior & Reinforcement Systems**: For tracking and encouraging positive behavior.

   ◇   **Communication & Family Engagement**: For seamlessly connecting with parents.

   ◇   **Routine & Task Automation**: For streamlining daily administrative tasks.

   ◇   **Classroom Organization & Noise Management**: For maintaining focus and order.

4   **Breakdown for Each App**: For every app, include:

   ◇   **App Name & Platform**: (e.g., "ClassDojo

(Web, iOS, Android)").

◇ **Category**: Label it from the list above (e.g., Behavior & Reinforcement).

◇ **Core Function**: A 1-2 sentence description of what it does.

◇ **The "Why" it's Important**: A brief explanation of the specific problem it solves and its long-term impact (e.g., "This app creates a strong positive reinforcement system that visually rewards specific behaviors, motivating students and providing parents with a real-time window into their child's day, which dramatically improves the home-school partnership.").

◇ **Ideal Use Case**: A specific example of how it is used in a classroom setting.

5 **The "Why" Behind This Strategy**: A concluding summary explaining how this multi-tool approach creates a cohesive management system that is greater than the sum of its parts. It reduces teacher burnout from administrative overload, creates a culture of positivity and transparency, and provides the data needed to make informed instructional and behavioral decisions.

6 **Bonus**: Pro-Tip for Teachers - Implementation & Privacy:

◇ **Start with One**: Don't try to implement all these at once. Master one app that addresses your biggest pain point first.

◇ **Student Ownership**: Whenever possible, allow students to interact with the app (e.g., giving themselves points for completing a task, checking off a to-do list).

◇ **Privacy First**: Always review an app's privacy policy (especially COPPA and FERPA compliance) and obtain any necessary school or parent permissions before using it with students.

◇ **The Power of a Single Platform**: Whenever possible, choose an app that combines multiple functions (e.g., communication + behavior) to avoid app fatigue for yourself and parents.

The final output should be a complete, plug-and-play digital management toolkit that empowers educators to work smarter, not harder.

Complete Output:

The "Why" it Matters: Effective classroom management is built on consistency, clarity, and communication. Technology apps excel at providing these things at scale. They turn abstract expectations into tangible systems, provide instant feedback, and open a vital communication channel with families. This allows the teacher to be a more present, proactive facilitator of learning rather than a full-time manager of logistics and behavior.

The App Menu: Your Digital Management Toolkit

1. App: ClassDojo (Web, iOS, Android)

- **Category**: Behavior & Reinforcement / Communication
- **Core Function**: A behavior-tracking platform where teachers can award positive (and negative) points to students for specific skills. It also features a built-in messaging system for parents and a story feed for sharing class photos.
- **Why It's Important**: It makes behavior management visual and positive. Students love customizing their av-

atars and earning feedback. The direct link to parents fosters a powerful team approach to supporting student behavior and celebrating successes.

- **Ideal Use Case**: Giving a point to the entire class for a smooth transition, or awarding a point to an individual student for perseverance. Sending a quick photo message to all parents about a fun science experiment.

2. App: Google Classroom (Web, iOS, Android)

- **Category**: Routine & Task Automation
- **Core Function**: A centralized hub for creating, distributing, and grading assignments. It seamlessly integrates with Google Docs, Slides, and Sheets.
- **Why It's Important**: This is the ultimate tool for streamlining workflow and going paperless. It eliminates the "I lost my handout" excuse, organizes all student work in one place, and makes providing digital feedback efficient. It automates the distribution and collection of work.
- **Ideal Use Case**: Posting a daily assignment with attached resources, creating a quiz that auto-grades, and using the "comment bank" feature to leave frequent feedback.

3. App: Classcraft (Web, iOS, Android)

- **Category**: Behavior & Reinforcement / Classroom Organization
- **Core Function**: Turns classroom management into a role-playing game (RPG). Students create avatars, earn points for positive behaviors, and work in teams. They can unlock "powers" (real-world rewards like extra time on a test).
- **Why It's Important**: It leverages gamification to create incredible buy-in, especially in upper elementary and

middle school. It teaches teamwork, responsibility, and consequences in a highly engaging format that feels like play.

- **Ideal Use Case**: Using the "random picker" to select students fairly, or having a student use a "power" to ask a team member for help during an assignment.

4. App: Remind (Web, iOS, Android)

- **Category**: Communication & Family Engagement
- **Core Function**: A safe, simple messaging app that allows teachers to send text messages to students and parents without sharing personal phone numbers. Supports translations.
- **Why It's Important**: It meets families where they are—on their phones. It's perfect for quick reminders about deadlines, field trips, or schedule changes. The translation feature is crucial for engaging multilingual families.
- **Ideal Use Case**: Sending a quick blast: "Reminder: Library books are due tomorrow!" or "Great job on the science fair today! Ask your child about their project."

5. App: Bouncy Balls (website) or Too Noisy (iOS/Android)

- **Category**: Classroom Organization & Noise Management
- **Core Function**: A fun noise monitor that uses a microphone to display the noise level in the room visually (e.g., bouncy balls jump higher, a needle moves into the "red" zone).
- **Why It's Important**: It provides an objective, visual cue for acceptable noise levels during collaborative work. It helps students self-monitor and regulate their volume, making them responsible for managing their

environment.

- **Ideal Use Case**: During group work or independent work time, project the tool on the board. Set the sensitivity with the class and challenge them to keep the balls calm.

6. App: Timer Tab (website) or built-in phone timer

- **Category**: Routine & Task Automation
- **Core Function**: A simple, large, and highly visible online or app-based timer and stopwatch.
- **Why It's Important**: It creates structure, manages pacing, and adds a sense of urgency to tasks. It helps both teachers and students manage time effectively and makes transitions between activities smooth and predictable.
- **Ideal Use Case**: Timing a 10-minute quick-write, using the stopwatch feature to time presentations, or giving a 5-minute warning for cleaning up.

The "Why" Behind This Strategy: This plan works because it addresses the core pillars of management: Behavior (ClassDojo/Classcraft), Communication (Remind), and Routine (Google Classroom/Timer). Using a combination of these tools creates a cohesive system where expectations are clear, feedback is immediate, and routines run smoothly. This integrated approach saves invaluable time, reduces stress, and creates a classroom environment where both students and teachers can thrive.

Bonus: Pro-Tip for Teachers - Implementation & Privacy:

- **Avoid Over-notification**: Be strategic with messaging apps like Remind. Too many messages will cause parents to mute them. Only send essential, high-value communications.

- **Co-create Behaviors**: When using ClassDojo or Class-craft, have the class help decide which behaviors earn points. This creates ownership and ensures the system is fair.

- **Tech is a Tool, Not a Total Solution**: These apps are most effective when paired with strong teacher-student relationships and well-established classroom routines. They enhance good practice; they don't replace it.

Prompt: Behavior Management Techniques

Simple Prompt: "Act as an expert in Educational Psychology and Positive Behavior Support. Your task is to create a comprehensive guide to behavior management techniques for a [GRADE LEVEL] classroom".

- **Tone**: Empathetic, Evidence-Based
- **Format**: Behavior Technique List (x5)
- **Platform**: Teacher Training, Classroom Guides
- **AI Role**: Behavioral Coach
- **Prompt Goal**: Support teachers with practical approaches for discipline
- **Tags**: #behaviorManagement, #classroomSupport, #teachingTools
- **Prompt Variables**: {grade/level}, {behaviorTypes}, {classSize}

Best Practice Tip: Use restorative practices alongside corrective methods to build long-term respect.

Enhanced Prompt:

Act as an expert in Educational Psychology and Positive Behavior Support. Your task is to create a comprehensive guide to behavior management techniques for a [GRADE LEVEL] classroom. The goal is to provide a toolkit of strategies that are proactive, restorative, and designed to teach self-regulation, preserve student dignity, and maintain a positive learning environment for all students.

The guide must be structured as follows:

1. **The "Why" it Matters**: A brief, impactful explanation of the core philosophy behind effective management (e.g., "Behavior management is not about control or compliance; it is about teaching. The goal is to instruct students on how to meet their needs appropriately within a social context. The most effective techniques are preventative, relationship-based, and focus on reinforcing desired behaviors rather than punishing unwanted ones, thereby creating a culture of accountability and respect.").

2. **Overarching Goal**: State the primary objective of this management philosophy (e.g., "To cultivate a classroom where students feel safe, connected, and capable, which is the foundation for academic risk-taking and social-emotional growth. We aim to build intrinsic motivation, not just extrinsic compliance.").

3. **The Strategic Toolkit**: A Tiered Approach Provide a list of 5-7 distinct techniques. The techniques must be categorized into a logical, tiered framework:

   ◇ **Tier 1**: Proactive & Preventative (For All Students): Techniques to design the environment and teach expectations to prevent most misbehavior.

   ◇ **Tier 2**: Responsive & Supportive (For Some Students): Techniques to provide subtle, non-disruptive cues and minimal interventions to redirect off-task behavior.

   ◇ **Tier 3**: Restorative & Reflective (For Individual Students): Techniques to address persistent behaviors, understand their root cause, and repair harm.

4. **Breakdown for Each Technique**: For every tech-

nique, include:

◇ **Technique Name**: A clear, actionable name.

◇ **Tier**: Label it as Proactive, Responsive, or Restorative.

◇ **How It Works**: A concise, step-by-step explanation of how to implement the technique in a classroom setting. Be specific and practical.

◇ **The "Why" it's Important**: A brief explanation of the psychological or pedagogical principle it leverages and its long-term impact on student behavior (e.g., "This technique uses non-verbal cues to correct behavior without public shame, preserving the student's relationship with the teacher and their peers while maintaining instructional momentum.").

5 **The "Why" Behind This Strategy**: A concluding summary explaining how this tiered, teaching-oriented approach is more sustainable and effective than a purely punitive model. It reduces power struggles, builds student self-awareness and executive functioning skills, and allows the teacher to act as a coach rather than a disciplinarian.

6 **Bonus**: Pro-Tip for Teachers - The Key to Implementation:

◇ **Focus on Tier 1**: "Invest 80% of your energy in perfecting your Tier 1 proactive strategies. A strong, predictable, and engaging classroom prevents 90% of behavioral issues."

◇ **The Pause Principle**: "When a behavior occurs, take a breath before reacting. Your calmness is de-escalating. Ask yourself: 'Is this a can't do or a won't do problem? Is the student lacking the skill or the motivation?'"

◇ **The Power of "I Notice"**: "Use language that describes the desired behavior you want to see rather than highlighting the misbehavior (e.g., 'I see table two is ready with their eyes on me' instead of 'Stop talking and look up here.')."

◇ **Re-teach Routines**: "Assume that all behaviors will need re-teaching, especially after breaks or during times of high stress. This is not a sign of failure but of effective instruction."

The final output should be a complete, plug-and-play management toolkit that empowers educators to build a respectful, self-regulated classroom community.

Complete Output:

The "Why" it Matters: Effective behavior management is the bedrock of a functional classroom. It is not about exerting control but about creating conditions where positive behaviors can flourish. By focusing on teaching expectations, reinforcing desired behaviors, and addressing misbehavior through a lens of learning, we build students' intrinsic motivation and social-emotional skills, which are far more valuable than simple compliance.

The Techniques: A Tiered Toolkit

1. Technique: Procedural Routines & Rehearsal (Tier 1: Proactive)

- **How It Works**: Design, explicitly teach, and practice specific routines for common transitions (e.g., entering the class, turning in work, group work). Practice them until they are automatic. Use cues like a timer or a specific phrase to initiate the routine.

- **Why It's Important**: Predictability reduces anxiety and confusion. Clear routines eliminate the "what do I do now?" moment that often leads to off-task behavior, saving valuable instructional time and creating a calm, orderly environment.

2. Technique: Non-Verbal Cues (The "Look" or Signal) (Tier 2: Responsive)

- **How It Works**: Establish a silent signal for common corrections (e.g., making eye contact and tapping your ear to remind a student to listen, pointing to the expected work on their desk). This is a private reminder between you and the student.

- **Why It's Important**: It addresses the behavior immediately and effectively without disrupting the lesson or drawing negative peer attention to the student, thus preserving their dignity and your teaching flow.

3. Technique: Proximity Praise (Tier 2: Responsive)

- **How It Works**: Instead of correcting the off-task student, specifically praise a nearby student who is demonstrating the desired behavior (e.g., "I love how Sarah has her book open to page 42 and is ready to discuss."). The off-task student will often self-correct to also receive praise.

- **Why It's Important**: This technique positively reinforces the correct behavior for the entire class to hear and provides a clear model to follow. It is a subtle yet powerful way to redirect without direct confrontation.

4. Technique: "I" Statements & Neutral Language (Tier 2: Responsive)

- **How It Works**: When addressing behavior, state the observable fact and its impact using neutral, non-accusatory language (e.g., "When there is talking while I'm giving instructions, I feel frustrated because students might miss important information and have to redo their work," instead of "Stop being rude and talking!").

- **Why It's Important**: This de-escalates conflict by removing blame and focusing on the collective impact of the behavior. It models respectful communication and helps students understand the "why" behind the rules.

5. Technique: Offer Choice (Tier 2: Responsive / Tier 3: Restorative)

- **How It Works**: When a student is resistant, provide two acceptable choices (e.g., "You can choose to work on the essay outline first or the research notes. Which feels more manageable to start with?" or "You can choose to rejoin the group respectfully or you can take five minutes to reset at the calm-down corner.").

- **Why It's Important**: Choice empowers students and gives them a sense of control, which can often defuse a power struggle. It teaches decision-making and problem-solving skills.

6. Technique: Restorative Chat (Tier 3: Restorative)

- **How It Works**: Privately, ask a series of non-blaming questions to understand the root cause and repair harm: "What happened?" "What were you thinking at the time?" "Who has been affected by your actions?" "What do you need to do to make things right?"

- **Why It's Important**: This moves beyond punishment to focus on accountability, empathy, and repairing relationships. It helps identify if the behavior stems from a lack of skill, an unmet need, or an emotional regulation issue.

7. Technique: Behavior-Specific Praise (Tier 1: Proactive)

◇ **How It Works**: Instead of saying "Good job," praise the specific action you want to see repeated (e.g., "Thank you for pushing in your chair so the aisle is clear for others," or "I noticed you helped David find the right page. That was very collaborative of you.").

◇ **Why It's Important**: This clearly communicates to the student—and the class—exactly what constitutes desired behavior. It reinforces the action and motivates other students to exhibit the same behavior.

The "Why" Behind This Strategy: This plan works because it shifts your role from a reactive disciplinarian to a proactive teacher of behavior. By investing in prevention and using subtle, respectful interventions, you build trust with students. This trust is the currency of a positive classroom culture, where students feel safe to learn and make mistakes, ultimately leading to more effective teaching and deeper learning for everyone.

Bonus: Pro-Tip for Teachers - The Key to Implementation:

- **Start Small**: Choose one technique to master for a week. Once it feels natural, add another.

- **Your Calm is Your Superpower**: Regulate your own emotions first. A deep breath and a neutral tone can de-escalate almost any situation.

- **Re-teach Constantly**: Assume procedures and expectations need to be re-taught after weekends, holidays, and anytime the classroom energy feels off. This is not a failure; it's good teaching.

- **Connect Before You Correct**: A positive, genuine relationship is the foundation upon which all these techniques are built. A student will rarely respond well to redirection from a teacher they feel doesn't like them.

Prompt: Time Management for Teachers

Simple Prompt: "Act as an expert in Educational Efficiency and Sustainable Teaching Practices. Your task is to create a comprehensive guide to time management strategies specifically for teachers".

- **Tone**: Practical, Structured
- **Format**: Weekly Time Plan
- **Platform**: Teacher Journals, Planner Apps
- **AI Role**: Productivity Consultant
- **Prompt Goal**: Help teachers manage workload effectively and reduce stress
- **Tags**: #timeManagement, #teacherSupport, #productivity
- **Prompt Variables**: {teachingLoad}, {availableTime}, {priorities}

Best Practice Tip: Use batching (grading, planning) to reduce context switching and save energy.

Enhanced Prompt:

Act as an expert in Educational Efficiency and Sustainable Teaching Practices. Your task is to create a comprehensive guide to time management strategies specifically for teachers. The goal is to provide a toolkit of strategies that protect planning time, maximize instructional impact, and prevent burnout by focusing on intentionality, prioritization, and systemization over sheer effort.

The guide must be structured as follows:

1. **The "Why" it Matters**: A brief, impactful explanation of the core philosophy behind teacher time management (e.g., "For teachers, time management is not a productivity hack; it is a sustainability strategy. It is the deliberate practice of aligning your finite time and energy with your highest-impact priorities—both in and out of the classroom. Effective time management reduces overwhelm, creates space for deep work and personal well-being, and is the foundational skill that prevents burnout and fosters a long, fulfilling career.").

2. **Overarching Goal**: State the primary objective of this time management philosophy (e.g., "To move from a reactive state of constant busyness to a proactive state of intentional teaching, where time is spent on tasks that directly impact student learning and teacher well-being, while minimizing energy drain on low-impact activities.").

3. **The Strategic Toolkit**: A Thematic Approach Provide a list of 5-7 distinct strategies. The strategies must be categorized into a logical framework that addresses different facets of a teacher's life:

   ◇ **Proactive Planning**: Strategies for intentional weekly/daily structuring.

   ◇ **In-Class Efficiency**: Strategies to maximize instructional time.

   ◇ **Grading & Feedback**: Strategies to manage the assessment load.

   ◇ **Boundary Setting**: Strategies to protect personal time and prevent burnout.

4. **Breakdown for Each Time Management Strate-**

gy: For every strategy, include:

- ◊ **Strategy Name**: A clear, actionable name.

- ◊ **Category**: Label it as Proactive Planning, In-Class Efficiency, Grading & Feedback, or Boundary Setting.

- ◊ **How to Implement It**: A concise, step-by-step explanation of how to execute the strategy.

- ◊ **The "Why" it's Important**: A brief explanation of the long-term impact on efficiency, job satisfaction, and teaching effectiveness (e.g., "This strategy creates a predictable rhythm for the week, reduces decision fatigue, and ensures high-priority tasks are guaranteed dedicated time, making you more focused and less stressed.").

5 **The "Why" Behind This Strategy**: A concluding summary explaining how this holistic approach to time management—focusing on systems over willpower—creates sustainable teaching habits. It empowers teachers to reclaim agency over their time, leading to greater job satisfaction, more impactful teaching, and a healthier work-life balance.

6 **Bonus**: Pro-Tip for Teachers - The Mindset Shift:

- ◊ **The "Good Enough" Principle**: Not every task deserves your best effort. Apply your 'A' game to lesson planning and feedback, but your 'B' or 'C' game to decorating bulletin boards or creating elaborate sign-up sheets. Perfectionism is the enemy of efficiency.

- ◊ **Energy Cycling**: Schedule demanding, high-focus tasks (like planning) for when your energy is highest (e.g., Monday morning). Schedule low-energy tasks (like photocopying) for when your energy dips (e.g., Friday afternoon).

◇ **The 2-Minute Rule**: If a task will take less than two minutes (e.g., answering a quick email, filing a paper), do it immediately. This prevents small tasks from piling up into a daunting mental load.

◇ **Batch Processing**: Group similar tasks together (e.g., power hour for grading, one afternoon for making all copies for the week) to minimize context-switching, which is a major drain on mental energy.

The final output should be a complete, plug-and-play time management toolkit that provides realistic, actionable strategies to help teachers work smarter, not harder.

Complete Output:

The "Why" it Matters: Teacher time is a finite resource constantly pulled in a dozen directions. Without a proactive system, the urgent will always drown out the important. Strategic time management is the practice of consciously designing your workflow to ensure your energy is invested in the tasks that yield the highest return for student learning and your own professional well-being, transforming a reactive job into a proactive vocation.

The Strategic Toolkit: Reclaiming Your Time

1. Strategy: The Weekly "Power Hour" (Proactive Planning)

• **How to Implement It**: Block out one hour, once a

week (e.g., Friday afternoon or Sunday evening), to map out the upcoming week. Review lesson plans, create a daily to-do list, and schedule your priorities into your calendar. Treat this appointment with yourself as non-negotiable.

- **Why It's Important**: This single hour of proactive work prevents countless hours of reactive scrambling during the week. It provides mental clarity, reduces Sunday-night anxiety, and ensures you enter Monday focused and prepared.

2. Strategy: Time-Blocking Your Planning Period (In-Class Efficiency)

- **How to Implement It**: Before your planning period starts, decide exactly what you will accomplish in that time. Use a timer. Example: "Minutes 1-15: Make copies for tomorrow. Minutes 16-30: Respond to urgent emails. Minutes 31-45: Plan the opener for Tuesday's lesson."

- **Why It's Important**: Without a plan, planning periods are easily lost to distractions, chats with colleagues, or task paralysis. Time-blocking transforms this precious time into a focused, productive sprint, ensuring essential tasks actually get done.

3. Strategy: The "Do Now" / Bell Ringer Routine (In-Class Efficiency)

- **How to Implement It**: Every lesson should start with a predictable, 5-minute task already displayed when students enter (e.g., a review question, a journal prompt, a vocabulary exercise). This routine allows you to take attendance, handle minor issues, and students to transition mentally without wasting a minute of instructional time.

- **Why It's Important**: This strategy maximizes every instructional minute and establishes a tone of purpose and productivity from the moment students walk in. It saves up to 25 minutes of learning time per week.

4. Strategy: Targeted Feedback & Grading Sprints (Grading & Feedback)

- **How to Implement It**: Stop grading everything. Use selective grading—only grade a few key assignments in depth. For others, use spot-checking, completion grades, or whole-class feedback. When you do grade, set a timer for a "grading sprint" (e.g., 25 minutes of focused grading, then a 5-minute break).

- **Why It's Important**: This prevents the endless grading trap that consumes weekends. It shifts the focus from assigning points to providing actionable feedback that students can use, which is far more impactful for learning and sustainable for you.

5. Strategy: Digital Communication & Office Hours (Boundary Setting)

- **How to Implement It**: Set clear boundaries for communication. Use an auto-responder that states your email response time (e.g., "I respond to emails within 24 hours on weekdays"). Consider setting "virtual office hours" one evening a week for student/parent questions, instead of being constantly on-call.

- **Why It's Important**: This protects your personal and family time from the constant intrusion of work communication. It manages expectations, reduces your anxiety, and models healthy digital citizenship for students and parents.

6. Strategy: The "To-Don't" List (Proactive Planning)

- **How to Implement It**: Actively identify and eliminate low-impact tasks that consume your time. Examples: "I will not decorate elaborate monthly bulletin boards,"

"I will not grade every single practice problem," "I will not attend optional meetings that don't have a clear agenda."

- **Why It's Important**: This is about strategic abandonment. By consciously deciding what not to do, you free up massive amounts of time and mental energy for the tasks that truly matter for student learning.

The "Why" Behind This Strategy: This plan works because it addresses time management as a systemic issue, not a personal failing. It provides structures—like time-blocking and the "Power Hour"—that automate decision-making and protect your focus. By working within these designed systems, you reduce the cognitive load of constant prioritization, making you a more effective, less stressed, and more present educator.

Bonus: Pro-Tip for Teachers - The Mindset Shift:

- **Protect Your Morning & Evening Routines**: Do not check email first thing in the morning or right before bed. This sets a reactive, anxious tone for the day or prevents restful sleep. Guard this time fiercely.

- **Delegate to Students**: Students can pass out papers, manage materials, and even help with simple classroom tasks. This teaches them responsibility and saves you precious minutes every day.

- **Reflect and Refine**: At the end of each week, spend 5 minutes reflecting: "What felt chaotic? What system can I put in place to prevent that next week?" Continuous small tweaks lead to massive long-term gains.

Prompt: Professional Development Goals

Simple Prompt: Act as an expert in Adult Learning and Educator Development. Your task is to create a comprehensive menu of professional development goals for teachers in [subject/role].

- **Tone**: Supportive, Forward-Looking
- **Format**: Growth Plan / Goal List
- **Platform**: Teacher Evaluations, PD Sessions
- **AI Role**: Teacher Coach
- **Prompt Goal**: Guide teachers in setting and achieving career development goals
- **Tags**: #professionalDevelopment, #teacherGrowth, #education
- **Prompt Variables**: {subject/role}, {experienceLevel}, {goals}

Best Practice Tip: Encourage teachers to set one short-term and one long-term goal for balance.

Enhanced Prompt:

Act as an expert in Adult Learning and Educator Development. Act as an expert in Adult Learning and Educator Development. Your task is to create a comprehensive menu of professional development goals for teachers in [subject/role]. The goal is to provide a framework for goals that are not just compliance-oriented checkboxes, but are transformative, personalized, and directly linked to enhancing student learning and teacher fulfillment. These goals should foster a mindset of continuous improvement and reflective practice.

The guide must be structured as follows:

1. **The "Why" it Matters**: A brief, impactful explanation of the core philosophy behind meaningful professional development (e.g., "Effective professional development is a career-long journey of refinement, not a one-time event. It is a personalized commitment to moving from being a good teacher to a masterful one. The best PD goals are self-directed, intrinsically motivated, and focused on a specific, high-leverage area of practice that, when improved, creates a ripple effect of positive outcomes for both the teacher and their students.").

2. **Overarching Vision**: State the primary objective of this growth-oriented philosophy (e.g., "To cultivate a professional practice that is continuously evolving, evidence-based, and deeply connected to the specific needs of one's students, leading to greater instructional impact, personal job satisfaction, and professional leadership.").

3. **The Goal Menu**: A Framework for Growth Provide a list of 5-7 distinct goal ideas. The goals must be categorized into a logical framework that addresses different dimensions of a teacher's practice:

   ◇ **Pedagogical Craft**: Goals focused on refining instructional techniques.

   ◇ **Content & Curriculum**: Goals focused on deepening knowledge and resource development.

   ◇ **Assessment & Data**: Goals focused on improving the use of evidence to inform teaching.

   ◇ **Community & Collaboration**: Goals focused on building professional relationships and leadership.

   ◇ **Professional Wellness**: Goals focused on sustainability and preventing burnout.

4   **Breakdown for Each Professional Development Goal**: For every goal, include:

◇   **Goal Title**: A clear, focused name for the goal.

◇   **Category**: Label it from the list above (e.g., Pedagogical Craft).

◇   **The SMART Goal**: A template written in the Specific, Measurable, Achievable, Relevant, and Time-bound format.

◇   **The "Why" it's Important**: A brief explanation of the long-term impact on teaching effectiveness and student outcomes (e.g., "Mastering a few high-yield instructional routines creates a more predictable and effective learning environment, increases student engagement, and allows the teacher to focus on nuanced student thinking rather than managing activity transitions.").

◇   **First Steps & Evidence of Success**: 1-2 concrete actions to begin and how to know if the goal is being met.

5   **The "Why" Behind This Strategy**: A concluding summary explaining how this self-directed, personalized approach to PD leads to more authentic and lasting change than top-down mandates. It empowers teachers to own their growth, aligns their development with their passions and their students' needs, and builds a culture of excellence and shared learning within a school.

6   **Bonus**: Pro-Tip for Teachers - Making it Stick:

◇   **Start Small**: Choose ONE goal per semester. Depth over breadth. Trying to change everything at once leads to changing nothing.

◇ **Find a Peer**: Partner with a colleague who has a similar goal. Observe each other, share resources, and provide non-evaluative feedback. This accountability partnership is a game-changer.

◇ **Document Your Journey**: Keep a simple log or journal. What did you try? What worked? What didn't? This becomes evidence of your growth and a valuable resource for your portfolio.

◇ **Share Your Learning**: Present a brief overview of your goal and findings at a department meeting. Teaching others consolidates your own learning and contributes to the school's professional capital.

The final output should be a complete, plug-and-play professional growth kit that empowers educators to design a personalized and impactful path for their own development.

Complete Output:

The "Why" it Matters: True professional development is an investment in your most valuable teaching asset: yourself. It moves beyond mandatory trainings to a deliberate practice of honing your craft. By focusing on a specific, self-chosen area of growth, you reclaim agency over your career, combat stagnation, and directly amplify your effectiveness and impact in the classroom. This is the path from burnout to mastery.

The Goal Menu: A Framework for Your Growth

1. Goal: Deepen a High-Yield Instructional Routine

- **Category**: Pedagogical Craft

- **The SMART Goal**: "This semester, I will implement and refine the 'Think-Pair-Share' strategy in my Grade 9 History classes at least three times per week. I will measure success through student engagement metrics and a reflection journal."

- **Why It's Important**: Mastering a few core routines like this creates a more dialogic classroom, ensures equitable participation, and provides formative assessment data. It builds a foundation of student-centered practice upon which other strategies can be built.

- **First Steps & Evidence**:

  ◇ **First Step**: Plan the routine into next week's lessons. Create clear sentence stems to support paired discussions.

  ◇ **Evidence of Success**: You notice less teacher-talk and more student-to-student discourse. Anecdotal records show more students are willing to share with the whole class after practicing with a partner.

2. Goal: Integrate Formative Assessment Techniques

- **Category**: Assessment & Data

- **The SMART Goal**: "By the end of the quarter, I will use at least one new formative assessment technique (e.g., exit tickets, Plickers, a Google Form quiz) in every unit to check for understanding and adjust my instruction in real-time."

- **Why It's Important**: This shifts assessment from an end-point judgment to a feedback loop for learning. It allows you to differentiate and re-teach before the summative test, leading to higher student mastery and fewer failures.

- **First Steps & Evidence**:

◇ **First Step**: Choose one tech tool or strategy to master. Use it in a low-stakes way.

◇ **Evidence of Success**: You cancel a planned lesson because the exit ticket data showed widespread confusion, and you re-teach the concept instead. Student performance on the subsequent summative assessment improves.

3. Goal: Develop a Peer Observation Habit

- **Category**: Community & Collaboration
- **The SMART Goal**: "This school year, I will conduct two peer observations: one with a teacher known for strong classroom management and one with a teacher known for project-based learning. I will follow each with a reflective conversation."
- **Why It's Important**: This breaks down teacher isolation and creates a culture of shared practice. It provides practical, contextualized ideas that are far more valuable than abstract theory and builds a supportive professional network.
- **First Steps & Evidence**:

  ◇ **First Step**: Identify two colleagues and respectfully ask to observe them. Schedule the time.

  ◇ **Evidence of Success**: You return to your classroom with one concrete, actionable strategy you can implement immediately. You build a new collegial relationship.

4. Goal: Curate a Digital Resource Library

- **Category**: Content & Curriculum
- **The SMART Goal**: "I will dedicate 30 minutes per week to finding, testing, and organizing high-quality digital resources for my Grade 6 Science units into a

curated Google Drive folder by the end of the semester."

- **Why It's Important**: This saves immense time in the long run. Having a vetted, easily accessible bank of resources (videos, simulations, articles) allows for faster lesson planning, easier differentiation, and more engaging instruction.

- **First Steps & Evidence**:

  - ◇ **First Step**: Create a Google Drive folder with subfolders for each unit. Add one resource you already use.

  - ◇ **Evidence of Success**: When planning a lesson, you can pull a relevant video or article in under two minutes instead of searching for 20.

5. Goal: Implement a Grading & Feedback System

- **Category**: Professional Wellness / Assessment

- **The SMART Goal**: "I will test and implement one strategy to reduce my grading time by 25% this semester (e.g., using more peer feedback, employing selective grading, using audio feedback, creating a single-point rubric)."

- **Why It's Important**: This goal is directly about sustainability. Reclaiming hours spent on grading reduces burnout, protects personal time, and can often lead to more meaningful, actionable feedback for students.

- **First Steps & Evidence**:

  - ◇ **First Step**: Audit your grading: what takes the most time? Research one alternative strategy.

  - ◇ **Evidence of Success**: You have your weekends back. Student feedback on assignments is more consistent and less delayed.

The "Why" Behind This Strategy: This plan works because it treats professional development as a personal and practical endeavor. By choosing goals that are specific, achievable, and directly relevant to your daily challenges, you ensure that your growth is authentic and impactful. This approach fosters intrinsic motivation, builds a portfolio of demonstrable skills, and ultimately leads to a more sustainable and rewarding teaching career.

Bonus: Pro-Tip for Teachers - Making it Stick:

- **Schedule It**: Block out non-negotiable time in your calendar for your PD goal work (e.g., "30 minutes on Tuesday for resource curation"). If it's not scheduled, it won't happen.

- **Celebrate Micro-Wins**: Acknowledged your progress! Successfully trying a new strategy once is a win. Share it with your accountability partner.

- **Be Kind to Yourself**: Growth is messy. Not every new strategy will work perfectly the first time. Reflect on what you learned from the attempt and adjust. The goal is progress, not perfection.

Prompt: Classroom Organization Tips

Simple Prompt: "Act as an expert in Educational Efficiency and Classroom Design. Your task is to create a comprehensive guide to classroom organization strategies for a [GRADE LEVEL] classroom".

- **Tone**: Practical, Organized
- **Format**: Tip List (x5)
- **Platform**: Classroom Guides, Teacher Blogs
- **AI Role**: Classroom Designer
- **Prompt Goal**: Improve classroom organization for smoother teaching
- **Tags**: #organization, #classroom, #teacherSupport
- **Prompt Variables**: {grade/subject}, {classSize}, {teachingStyle}

Best Practice Tip: Involve students in maintaining organization to build responsibility.

Enhanced Prompt:

Act as an expert in Educational Efficiency and Classroom Design. Your task is to create a comprehensive guide to classroom organization strategies for a [GRADE LEVEL] classroom. The goal is to provide a toolkit of tips that go beyond mere tidiness to create a highly functional, efficient, and student-centered environment that minimizes distractions, maximizes instructional time, and fosters student independence.

The guide must be structured as follows:

1. **The "Why" it Matters**: A brief, impactful explanation of the core philosophy behind classroom organization (e.g., "Classroom organization is the silent engine of effective teaching. A well-organized space is a proactive behavior management system. It reduces student anxiety, eliminates countless daily distractions ("Where's my pencil?"), and creates clear pathways for learning, allowing both the teacher and students to focus their cognitive energy on content and connection, not on clutter and chaos.").

2. **Overarching Goal**: State the primary objective of this organizational philosophy (e.g., "To design a classroom environment where physical space, materials, and workflows are intuitively structured, enabling students to independently access what they need, reducing clutter and transition time, and fostering a calm, efficient learning community").

3. **The Organizational Toolkit**: A Tiered Approach Provide a list of 5-7 distinct organizational tips. The tips must be categorized into a logical framework:

    ◇ **Physical Space & Zones**: Structuring the room's layout for specific purposes.

    ◇ **Systems & Procedures**: Creating routines for materials and workflow.

    ◇ **Student Empowerment**: Strategies to make students partners in organization.

4. **Breakdown for Each Organizational Tip**: For every tip, include:

    ◇ **Tip Title**: A clear, actionable name.

    ◇ **Category**: Label it as Physical Space, Systems, or Student Empowerment.

◇ **How to Implement It**: A concise, step-by-step explanation of how to set it up.

◇ **The "Why" it's Important**: A brief explanation of the long-term impact on efficiency, student behavior, and teacher sanity (e.g., "This system eliminates the 'I don't have a pencil' bottleneck that wastes the first 5 minutes of every class, teaching students responsibility and saving valuable instructional time.").

5 **The "Why" Behind This Strategy**: A concluding summary explaining how this intentional, system-based approach to organization is a form of self-care for the teacher and a foundational lesson in executive functioning for students. It creates a calm, predictable environment that reduces stress for everyone and makes the classroom a more effective place for teaching and learning.

6 **Bonus**: Pro-Tip for Teachers - The Maintenance Mindset:

◇ **The 5-Minute Reset**: Dedicate the last 5 minutes of the day or period for students to reset the classroom. It's a non-negotiable investment in tomorrow's sanity.

◇ **One-In, One-Out Rule**: Periodically audit supplies. Before new materials come in, encourage students to purge old, broken, or unused items.

◇ **Label Everything**: A simple label maker is your best friend. Labels remove ambiguity and tell students exactly where things belong, empowering them to help.

◇ **Start Small**: Don't try to reorganize your entire room in a day. Pick one area (e.g., the teacher desk, the supply corner) to master each week.

The final output should be a complete, plug-and-play organizational toolkit that empowers educators to design a classroom that runs itself, saving time and reducing stress all year long.

Complete Output:

The "Why" it Matters: Organization in the classroom is not about aesthetics; it's about cognitive load. Every time a student has to search for a material or a teacher has to dig for a handout, valuable mental energy and instructional time are wasted. A strategically organized classroom removes hundreds of these tiny obstacles, creating a seamless flow that allows everyone to focus on what truly matters: learning. It is the ultimate proactive strategy for a calm and productive classroom.

The Organizational Toolkit: Building a Self-Running Classroom

1. Tip: Define Clear Zones (Physical Space)

- **How to Implement It**: Designate specific areas of your room for specific purposes: a quiet reading nook, a group work zone with whiteboards, a teacher-led small group area, a supply station, and a designated "turn-in" spot.

- **Why It's Important**: Zones create physical cues for behavior and activity. Students understand the purpose of each space, which minimizes off-task behavior and smooths transitions between activities. It makes the room's workflow intuitive.

2. Tip: The "Must-Do, May-Do" System for Absent Work (Systems)

- **How to Implement It**: Have a clearly labeled hanging

file folder or bin for each day of the week. At the end of the day, place any handouts from that day's lesson in the corresponding folder. A student who was absent knows to check the folder for the day(s) they missed.

- **Why It's Important**: This eliminates the frantic morning scramble of "What did I miss?" and empowers students to take responsibility for catching up. It saves you time and ensures students don't fall behind due to absence.

3. Tip: Student Supply Kits (Student Empowerment)

- **How to Implement It**: Instead of a communal supply bin that becomes a mess, create individual supply kits for each student. Use zip-up pouches or small bins containing essentials like pencils, pens, a glue stick, and a highlighter. Store them in desk caddies or a designated shelf.

- **Why It's Important**: This eliminates arguments over supplies and time lost distributing materials. It teaches students to manage and be responsible for their own tools. A quick "kit check" at the end of the period ensures everything is accounted for.

4. Tip: The "One Stop Shop" Turn-In System (Systems)

- **How to Implement It**: Create a single, clearly labeled turn-in station with a separate tray or folder for each subject or class period. Teach the procedure for turning in work until it becomes automatic.

- **Why It's Important**: This prevents the tragedy of the lost homework on your desk. You always know where to look for completed work, and students have a clear, consistent procedure to follow. It streamlines grading and accountability.

5. Tip: "When I'm Done" Anchor Chart (Systems)

- **How to Implement It**: Co-create and post an anchor chart that lists 3-5 purposeful activities students can do when they finish their work early (e.g., "Read your independent reading book," "Practice this week's vocabulary on Quizlet," "Work on a bonus challenge problem").

- **Why It's Important**: This prevents early finishers from becoming a distraction. It promotes continuous learning, fosters independence, and differentiates instruction by providing built-in extension activities.

6. Tip: Delegate with Classroom Jobs (Student Empowerment)

- **How to Implement It**: Assign specific, rotating jobs to students (e.g., Materials Manager, Technology Chief, Librarian, Floor Monitor). Create a job chart with clear descriptions of responsibilities.

- **Why It's Important**: This builds a classroom community where everyone contributes. It saves you countless small tasks and gives students a sense of purpose and ownership over their environment. It also teaches real-world responsibility.

The "Why" Behind This Strategy: This plan works because it shifts organization from a solitary, teacher-driven chore to a shared, classroom-wide system. By investing time upfront to establish these structures and empower students, you create a self-sustaining environment that drastically reduces your mental and physical workload throughout the year. This leads to less teacher burnout, more student autonomy, and a classroom culture of order, respect, and efficiency.

Bonus: Pro-Tip for Teachers - The Maintenance Mindset:

- **The Weekly 5-Minute Purge**: Once a week, trigger a "5-minute purge" where everyone cleans out their desks, folders, and the common areas. This prevents clutter from accumulating.

- **Everything Has a Home**: If you find an item out of place, don't just put it away; ask, "Does this have a designated home? If not, where should it live?" This ensures your systems evolve with your needs.

- **Model It**: Your organization sets the tone. If your desk is a disaster zone, it subtly gives students permission to be disorganized. Model the systems you expect them to use.

Prompt: Classroom Icebreaker Questions

Simple Prompt: "Act as an expert in Social-Emotional Learning (SEL) and Educational Pedagogy. Your task is to create a comprehensive list of icebreaker questions for a [GRADE LEVEL/COURSE] classroom".

- **Tone**: Friendly, Approachable
- **Format**: Question List (x10)
- **Platform**: First-Day Activities, Group Discussions
- **AI Role**: Student Engagement Coach
- **Prompt Goal**: Build trust and community at the start of classes
- **Tags**: #icebreakers, #studentEngagement, #teaching
- **Prompt Variables**: {grade/level}, {subject}, {classSize}

Best Practice Tip: Use a mix of personal, fun, and topic-related questions to create connection.

Enhanced Prompt:

Act as an expert in Social-Emotional Learning (SEL) and Educational Pedagogy. Your task is to create a comprehensive list of icebreaker questions for a [GRADE LEVEL/COURSE] classroom. The goal is to generate questions that are strategically designed to do more than just break the silence; they must build genuine connections, establish psychological safety, activate prior knowledge, and set the tone for a collaborative and intellectually curious learning environment.

The guide must be structured as follows:

1 **The "Why" it Matters**: A brief, impactful explanation of the core principle behind intentional icebreakers (e.g., "The right icebreaker is a foundational investment in your classroom community. It is a high-leverage strategy to quickly build trust, normalize vulnerability, and signal to students that this is a space where their voice is heard and their experiences are valued. This initial investment pays dividends in increased participation, collaboration, and risk-taking throughout the entire year.").

2 **Overarching Goal**: State the primary objective of this icebreaker session (e.g., "To move beyond superficial introductions and foster authentic connections, activate students' cognitive and social-emotional readiness for learning, and collaboratively establish norms of respect and active listening.").

3 **The Icebreaker Menu**: A Strategic Mix Provide a list of 5-7 distinct questions. Ensure there is a deliberate mix of the following types to serve different purposes:

◇ **Community & Connection Builders**: To find common ground and build empathy.

◇ **Metacognitive & Reflective**: To get students thinking about their own thinking and learning.

◇ **Content-Based & Skill-Revealing**: To subtly introduce themes of the course and assess prior knowledge.

◇ **Creative & Imaginative**: To spark joy, lower inhibitions, and engage different parts of the brain.

4 **Breakdown for Each Icebreaker Question**: For

every question, include:

◇ **The Question**: Phrased precisely as the teacher should ask it.

◇ **Icebreaker Type**: Categorize it from the list above (e.g., "Creative & Imaginative").

◇ **The "Why" it's Important**: A brief explanation of the specific social-emotional or academic skill it targets and its long-term impact (e.g., "This question requires students to make connections and synthesize information, a key critical thinking skill. It also reveals diverse perspectives and normalizes that there is no single 'right' answer in this classroom.").

◇ **Implementation Tip**: Clear, practical advice for the teacher on how to facilitate it (e.g., "Use a Turn-and-Talk protocol first to ensure all students have time to think," "Model your answer first to provide a scaffold and build vulnerability.").

5 **The "Why" Behind This Strategy**: A concluding summary explaining how this intentional approach to icebreakers accelerates the formation of a cohesive learning community, provides the teacher with invaluable insights into student personalities and readiness, and creates a classroom culture where students feel safe to engage deeply with content and with each other.

6 **Bonus**: Pro-Tip for Teachers - Facilitation for Safety & Inclusion:

◇ **The Right to Pass**: Always state that students have the "right to pass" if they are not comfortable answering a question. This builds trust.

◇ **Teacher Participation**: Always answer the question yourself first. This models vulnerability, length of response, and tone.

◇ **Think-Pair-Share**: For deeper questions, use this structure. This gives everyone a low-risk chance to speak before sharing with the whole group.

◇ **Debrief the Process**: Afterward, ask, "What did we learn about our class from this activity?" This metacognitive step solidifies the community building.

The final output should be a complete, plug-and-play icebreaker kit that provides a clear path from awkward silence to a connected, engaged classroom ready to learn.

Complete Output:

The "Why" it Matters: The first few minutes of a class set the emotional and intellectual tone for the entire year. A well-chosen icebreaker is a diagnostic tool, a community-building exercise, and a values-setting activity all in one. It tells students they are seen as whole people, not just repositories for information, which is the first step toward creating a truly inclusive learning environment.

The Icebreaker Menu: Questions with Purpose

1. Question: "If you were to create a museum exhibit about one thing you're passionate about, what would it be and what would be the first item in the exhibit?"

- **Type**: Creative & Imaginative / Community Builder
- **Why It's Important**: This question moves beyond "what's your hobby?" to uncover deeper passions and

values. It allows students to showcase expertise in a non-academic area, building confidence and allowing peers to see them in a new light. It also taps into creativity and descriptive language.

- **Implementation Tip**: Model this with your own passion (e.g., "Mine would be on vintage typewriters, and the first item would be my great-grandfather's 1920 Underwood"). Use Think-Pair-Share to let everyone brainstorm before sharing with the whole group.

2. Question: "What is a 'superpower' you have that helps you learn? (It could be anything: asking great questions, making cool diagrams, being a great listener, etc.)"

- **Type**: Metacognitive & Reflective
- **Why It's Important**: This frames learning strengths as superpowers, which is empowering and positive. It forces metacognitive reflection from day one and begins to create a shared vocabulary around learning strategies. It also allows you to quickly identify potential collaborators and understand the diverse strengths in your room.

- **Implementation Tip**: Chart these on the board as students share. This visually validates their strengths and shows the collective power of the classroom.

3. Question: "What is one question you have about [COURSE SUBJECT] that you've always wanted to answer?" (e.g., "about space," "about history," "about how stories are written")

- **Type**: Content-Based & Skill-Revealing
- **Why It's Important**: This instantly frames your subject as a field of active inquiry, not a static set of facts. It activates prior knowledge and curiosity, and it provides you with a goldmine of information about student inter-

ests, misconceptions, and readiness that you can use to tailor your units.

- **Implementation Tip**: Record these questions on an anchor chart titled "Our Investigating Board" and return to them throughout the year as they are answered.

4. Question: "Complete this sentence: 'A perfect learning environment is one where...'"

- **Type**: Community Builder / Metacognitive

- **Why It's Important**: This is a covert way to collaboratively set classroom norms and expectations. Students are literally defining the culture they want to be a part of. It gives them agency and ownership over the classroom environment from the very beginning.

- **Implementation Tip**: Chart the responses and use them to co-create your class's list of norms or agreements. For example, if a student says "...where it's okay to make mistakes," you can propose that as a core norm.

5. Question: "What is one small thing that always makes your day better?"

- ◊ **Type**: Community & Connection Builder

- ◊ **Why It's Important**: This question is simple, safe, and universally accessible. It focuses on small joys, fostering positivity and empathy. It often reveals commonalities (e.g., "a sunny day," "my dog," "a good snack") that quickly build a sense of shared experience and community.

- ◊ **Implementation Tip**: This is a great quick warm-up or closing question. It's low-stakes and ends things on a positive note.

The "Why" Behind This Strategy: This plan works because it treats community building as academic work. These questions are designed to elicit responses that are inherently interesting and revealing, allowing students to connect on a human level while simultaneously practicing the kinds of thinking—metacognition, inquiry, creativity—that you will value all year long. This seamless integration ensures the activity feels purposeful, not perfunctory.

Bonus: Pro-Tip for Teachers - Facilitation for Safety & Inclusion:

- **Never Force Whole-Class Sharing**: Use small groups or pairs (Turn-and-Talk) for the initial response. Then, you can ask, "Who heard something fascinating from their partner they'd like to share?" This takes the pressure off and encourages active listening.

- **Be Mindful of Trauma**: Avoid questions that could inadvertently trigger painful memories (e.g., "What was the best vacation you ever took?" might be difficult for some students). Stick to questions focused on the present, preferences, and learning.

- **Use a Talking Stick**: For whole-group sharing, use an object to signify whose turn it is to speak. This manages conversation and reinforces respectful listening.

Prompt: Daily Growth Blueprint (Tchr Ver.)

Simple Prompt: "Act as a 'Personal Coach for Educators' and 'Institutional Wellness Architect'. Create a daily productivity blueprint for teachers to balance teaching, grading, planning, and self-care".

- **Tone**: Supportive, Structured
- **Format**: Daily Growth Plan
- **Platform**: Teacher Planner, Journals
- **AI Role**: Personal Coach
- **Prompt Goal**: Support teacher well-being and daily efficiency
- **Tags**: #teacherGrowth, #dailyPlan, #productivity
- **Prompt Variables**: {teachingLoad}, {timeAvailable}, {selfCareNeeds}

Best Practice Tip: Schedule short wellness breaks between heavy tasks to avoid burnout.

Enhanced Prompt:

You are a 'Personal Coach for Educators' and 'Institutional Wellness Architect'. Your expertise lies in applying principles from positive psychology, energy management, and organizational systems theory to the unique, high-demand environment of teaching. You design routines that protect a teacher's well-being as the non-negotiable foundation for effective instruction and student connection.

Primary Objective: To design a comprehensive, realistic, and sustainable Daily Operating System for a teacher. This system must create clear boundaries, maximize efficiency during contract hours, and fiercely protect time for recovery and self-care to prevent burnout and promote long-term career joy.

Core Requirements for the Blueprint: The system must be a holistic, multi-phase plan that addresses the entire arc of a teacher's day.

1. The Morning Anchor Routine (30-Minute Launch Sequence):

- Design a precise, ritual-based routine to be executed between arrival at school and the first bell.

- **Must Include**:

  - ◇ **A Cognitive Priming Task**: A 5-minute activity to set the day's intention (e.g., reviewing the one most important task for the day).

  - ◇ **A Calming Ritual**: A specific practice to manage pre-day anxiety (e.g., 3 minutes of mindful breathing while the classroom is still quiet).

  - ◇ **A Logistical Setup**: A quick, efficient physical setup of the classroom to reduce morning friction.

- **The Rule**: This time is for preparation, not creation. No lesson planning allowed.

2. The Workday Flow Strategy (Energy-Based Task Management):

- **Provide a clear framework for managing the four core domains**: Teaching, Grading, Planning, and Interruptions.

- **Batching Protocol**: Specify which tasks must be batched together and when (e.g., "All non-urgent email is processed only during two designated 15-minute windows: mid-morning and late afternoon.").

- **The "Moment of Intervention" Plan**: Offer a simple, repeatable script for handling unexpected interruptions (e.g., a colleague popping in) that respectfully protects focus time without damaging relationships.

- **The Power-Down Transition**: A 5-minute end-of-contract-hours ritual to "shut down" the workday physically and digitally (e.g., clear desk, write down the first task for tomorrow, close all tabs and email).

3. The After-School Decompression Ritual (30-Minute Mental Transition):

- Design a deliberate ritual to facilitate the psychological transition from "Teacher" to "Person."

- This must be an activity that is actively engaging and unrelated to work.

- **Examples**: A brisk walk without a phone, a short workout, listening to an audiobook/podcast for pleasure, a mindfulness app session.

- **The Rule**: No discussing school during this ritual.

4. The Non-Negotiable Self-Care Check-In (The Daily Maintenance):

- Prescribe one specific, sub-10-minute self-care activity that serves as a daily diagnostic.

- **Frame it as a check-in, not a chore. (e.g., "2-Minute Journal Prompt**: 'What drained my energy today? What gave me energy?'" or "5 minutes of guided stretching while focusing on releasing the day's tension.").

5. The Philosophical Foundation ("The Why"):

- **Clearly articulate the core principle behind this system. (e.g., "This system is built on the** principle of 'Designing Defaults.' By making the desired behaviors easy and the undesired ones hard, we conserve willpower for where it's needed most: in the classroom with students.").

6. The Burnout Preemption Protocol ("Next-Level Strategy"):

- Provide a specific strategy for adapting this system during known high-stress periods (e.g., finals, report cards, conferences).

- **This must be a pre-emptive, not reactive, plan. (e.g., "The 'Simplify to Survive' Rule**: One week before a high-stress period, pre-plan and freeze 3 days' worth of low-lift, high-impact review activities. This eliminates decision fatigue during the stressful week itself.").

Final Output Structure:

I require the final blueprint to be presented as a clear, compassionate, and easy-to-implement guide.

Output Format:

- **Title**: The Teacher's Sustainable Daily Operating System

- **Guiding Mantra**: A one-sentence philosophy (e.g., "Protect your energy to protect your practice.").

- **The Daily Blueprint**: A visually clean, time-based table outlining the four phases of the system (Morning Anchor, Workday Flow, After-School Decompression, Self-Care Check-In).

- **Columns**: | Time/Phase | Core Activity | Specific Actions | Key Rule |

- **Deep Dive**: Detailed explanations for each of the six core components listed above.

- **Pro-Tip**: One piece of practical advice for implementation (e.g., "For the first week, focus only on implementing the Morning Anchor and After-School Decompression rituals. Add the other components once those feel solid.").

Final Request:

Please generate the complete Teacher's Sustainable Daily Operating System blueprint.

Prompt: Educational Website Generator

Simple Prompt: "Act as an expert in EdTech Pedagogy and Digital Literacy. Your task is to create a curated list of educational websites for [SUBJECT/TOPIC] designed for [GRADE LEVEL]".

- **Tone**: Innovative, Practical
- **Format**: Website Idea List (x3)
- **Platform**: Edtech Projects, Teacher Resource Sharing
- **AI Role**: Edtech Strategist
- **Prompt Goal**: Inspire creation of educational websites for teaching and learning
- **Tags**: #websiteIdeas, #edtech, #educationInnovation
- **Prompt Variables**: {subject/skill}, {audience}, {learningGoals}

Best Practice Tip: Incorporate interactive elements like quizzes or forums to increase engagement.

Enhanced Prompt:

Act as an expert in EdTech Pedagogy and Digital Literacy. Your task is to create a curated list of educational websites for [SUBJECT/TOPIC] designed for [GRADE LEVEL]. The goal is to move beyond a basic list of websites to provide a strategic evaluation of online tools that are engaging, effective, and aligned with specific learning objectives. The focus is on websites that transform passive consumption into active learning, creation, and critical thinking.

The guide must be structured as follows:

1 **The "Why" it Matters**: A brief, impactful explanation of the core principle behind strategic technology integration (e.g., "In a digital age, the question is not if we use technology, but how. The right educational website acts as a force multiplier: it can provide personalized practice at scale, unlock new forms of creative expression, simulate complex systems, and connect students with primary sources and global perspectives. This intentional selection moves technology from being a 'babysitter' to a 'bridge' for deeper understanding.").

2 **Overarching Learning Objectives**: State the 2-3 primary skills or standards this technology is designed to support (e.g., "To provide differentiated skill practice," "To facilitate collaborative project-based learning," "To enable students to create and share multimodal demonstrations of understanding.").

3 **The Website Menu**: A Strategic Mix Provide a list of 5-7 distinct websites. Ensure there is a deliberate mix of the following types to serve different pedagogical purposes:

◇ **Skill & Drill**: For adaptive practice and fluency building.

◇ **Content Creation & Curation**: For students to demonstrate understanding through multimedia.

◇ **Simulation & Exploration**: For interacting with complex systems or concepts.

◇ **Collaboration & Communication**: For fostering teamwork and discussion.

◇ **Formative Assessment & Feedback**: For

providing real-time data on student understanding.

4   **Breakdown for Each Website**: For every website, include:

  ◇ **Website Name & Link**: The official name and URL.

  ◇ **Category**: Label it from the list above (e.g., Simulation & Exploration).

  ◇ **Core Function**: A 1-2 sentence description of what it does.

  ◇ **The "Why" it's Important**: A brief explanation of the specific learning benefit and long-term impact (e.g., "This tool provides a safe, cost-effective way to experiment with complex scientific concepts, allowing for trial and error without real-world consequences and building deep conceptual understanding through inquiry.").

  ◇ **Ideal Use Case**: A specific example of when and how to integrate it into a lesson or unit (e.g., "Perfect for introducing a new physics concept, or for allowing students to form and test their own hypotheses in a virtual lab.").

5   **The "Why" Behind This Strategy**: A concluding summary explaining how this curated, purpose-driven approach to technology prevents "app fatigue," ensures screen time is meaningful and productive, and empowers teachers to make informed decisions that directly support their instructional goals.

6   **Bonus**: Pro-Tip for Teachers - The Integration Framework:

◇ **The SAMR Model**: Use this as a filter: Does the website simply Substitute a worksheet (okay), or does it Augment, Modify, or Redefine the learning task (ideal)?

◇ **Data-Driven Decisions**: Choose websites that provide you with actionable data on student progress, not just a final score.

◇ **Balance is Key**: Technology is one tool in the toolbox. The best learning environments blend tech-enabled activities with hands-on, unplugged, and social learning.

◇ **Student Choice**: Where possible, offer a "menu" of websites for students to choose from to demonstrate their understanding, fostering ownership and creativity.

The final output should be a complete, plug-and-play EdTech integration guide that is easy to personalize and provides clear, curriculum-aligned value.

Complete Output:

The "Why" it Matters: A strategically chosen educational website is a gateway to personalized and limitless learning. It can provide a struggling student with the customized practice they need to achieve mastery, while simultaneously challenging an advanced learner with complex, open-ended problems. This moves the classroom from a one-pace-fits-all model to a dynamic, differentiated learning environment.

The Website Menu: A Strategic Toolkit for [SUBJECT]

1. Website: Khan Academy (khanacademy.org)

- **Category**: Skill & Drill / Tutorial
- **Core Function**: Provides a vast library of practice ex-

ercises, instructional videos, and a personalized learning dashboard across numerous subjects.

- **Why It's Important**: It offers mastery-based learning, allowing students to work at their own pace to fill gaps in understanding or accelerate ahead. It provides teachers with unparalleled data on individual and class-wide progress.

- **Ideal Use Case**: Station rotation for differentiated practice, flipped classroom model, or targeted intervention for struggling students.

2. Website: PhET Interactive Simulations (phet.colorado.edu)

- **Category**: Simulation & Exploration

- **Core Function**: Provides free, research-based science and math simulations from the University of Colorado Boulder.

- **Why It's Important**: It makes abstract, invisible, or dangerous concepts (like gravity, circuit building, or natural selection) tangible, visual, and safe for students to manipulate and explore. This builds deep conceptual understanding through inquiry.

- **Ideal Use Case**: Introducing a new physics concept, conducting virtual labs where physical materials are unavailable, allowing students to form and test hypotheses.

3. Website: Canva for Education (canva.com/education/)

- **Category**: Content Creation & Curation

- **Core Function**: A simple, powerful graphic design tool with thousands of templates for presentations, infographics, posters, and videos.

- **Why It's Important**: It allows students to demonstrate their understanding visually and creatively, moving beyond the traditional essay or report. This fosters digital literacy and design thinking skills highly relevant to the modern world.

- **Ideal Use Case**: Creating science fair posters, designing book covers for a novel study, producing public service announcements, making data visualizations.

4. Website: Flip (info.flip.com)

- **Category**: Collaboration & Communication / Formative Assessment
- **Core Function**: A video discussion platform where teachers post prompts and students respond with short videos.
- **Why It's Important**: It amplifies student voice, especially for shy or reluctant writers. It builds a community of learners and allows for authentic, verbal demonstration of understanding. It also helps develop communication and digital citizenship skills.
- **Ideal Use Case**: Exit tickets, book talks, peer feedback sessions, explaining problem-solving steps orally.

5. Website: Quizizz (quizizz.com)

- **Category**: Formative Assessment & Feedback
- **Core Function**: A game-based platform for conducting formative assessments, quizzes, and interactive lessons. Features memes and music to increase engagement.
- **Why It's Important**: It provides instant, whole-class data on student understanding in a highly engaging, low-stakes format. The self-paced mode is perfect for differentiation, and the detailed reports help teachers identify learning gaps immediately.
- **Ideal Use Case**: Unit reviews, quick checks for understanding, bell ringers, and homework that students actually want to do.

6. Website: Google Jamboard (jamboard.google.com)

- **Category**: Collaboration & Communication
- **Core Function**: A cloud-based, collaborative digital whiteboard.
- **Why It's Important**: It facilitates real-time, visual collaboration. Students can brainstorm, map ideas, and problem-solve together from any device, capturing the energy of a group whiteboard session in a digital, savable format.
- **Ideal Use Case**: Brainstorming sessions, Venn diagrams, mind mapping, group problem-solving in math, "gallery walks" to view each other's work.

The "Why" Behind This Strategy: This plan works because it treats technology as a strategic partner, not a novelty. By carefully matching the website's function to a specific learning goal, you ensure that technology use is always intentional and effective. This curated approach saves teachers from the overwhelm of the app store and provides a versatile toolkit to support differentiation, engagement, and authentic assessment throughout a unit.

Bonus: Pro-Tip for Teachers - The Integration Framework:

- **SAMR in Action**: Ask: Can this tool Redefine the task? Instead of a poster (Substitution), could students use Canva to create an infographic that is published online for a real audience (Modification/Redefinition)?
- **Manage the Workflow**: Use a platform like Google Classroom or Seesaw as a "home base" to distribute links and collect student work from various websites. This simplifies the process for everyone.
- **Always Have a Non-Tech Backup**: Technology can fail. Always have an analogous unplugged activity ready to go to avoid losing instructional momentum.

Category 5: Student Growth & Development

Theme:

Overview of Student Growth & Development

This category is dedicated to fostering both personal and academic growth among students. Its focus is on equipping learners with the tools and strategies necessary to build confidence, develop resilience, enhance critical thinking, and establish healthy habits. By integrating resources for learning, mindset development, and well-being, this category ensures that students receive support in all areas of their development. The aim is to create a holistic environment where every student has the opportunity to succeed both in and out of the classroom.

Category Goal:

Empowering Lifelong Skills for Students and Educators

This framework is designed to provide both students and educators with practical tools and strategies that foster the development of essential lifelong skills. The focus is on nurturing capabilities such as critical thinking, effective research, resilience, personal health, and reflective practices. By emphasizing critical thinking and research strategies, learners are equipped to analyze information, solve problems creatively, and make informed decisions. These skills support academic achievement and prepare students to navigate complex challenges beyond the classroom.

The framework prioritizes resilience and health, encouraging students to cultivate habits that promote well-being. Building resilience helps learners adapt to setbacks, manage stress, and persevere through difficulties, ensuring they remain motivated and engaged. Reflective practices are integrated to help students assess their progress, set meaningful goals, and continually improve. Reflection fosters self-awareness and a growth mindset, empowering students to take ownership of their learning and personal development.

Through these interconnected approaches, learners are empowered to thrive not only within the academic environment but also in their broader lives, ensuring personal and educational success.

Mini-Index: Student Growth and Development

Prompt	AI Role	Prompt Goal
Book/Article Summary Generator	Literary Summarizer	Condense texts into student-friendly summaries
Essay Prompt Generator	Academic Coach	Inspire critical essays and structured writing
Sample Research Paper Generator	Research Mentor	Provide outlines and introductions for papers
Research Topic Generator	Topic Advisor	Suggest relevant and feasible research topics
Critical Thinking Exercise	Critical Thinking Mentor	Develop analytical and evaluative skills
Differentiated Instruction Strategies	Inclusive Educator	Adapt lessons for varied learning needs
Visualization Roadmap (Student)	Visualization Guide	Support mindset growth through guided visualization
Emotional Resilience Builder	Resilience Coach	Strengthen coping skills and emotional resilience
Growth Tracking Journal	Journaling Mentor	Encourage reflection and progress tracking
Health Habit Tracker	Wellness Coach	Support student well-being with habit tracking

Detailed Prompts

Prompt: Book/Article Summary Generator

Simple Prompt: "Act as an expert Literary Analyst and Pedagogical Specialist. Your task is to create a comprehensive summary and analysis guide for [BOOK/ARTICLE/PLAY/POEM] tailored specifically for [GRADE LEVEL]".

- **Tone**: Clear, Concise
- **Format**: Summary (1 page)
- **Platform**: Study Guides, Assignments, Student Notes
- **AI Role**: Literary Summarizer
- **Prompt Goal**: Help students quickly grasp the key points of texts
- **Tags**: #bookSummary, #studyAid, #studentLearning
- **Prompt Variables**: {book/article}, {themes}, {grade/level}

Best Practice Tip: Encourage students to compare the AI summary with their own notes to deepen comprehension.

Enhanced Prompt:

Act as an expert Literary Analyst and Pedagogical Specialist. Your task is to create a comprehensive summary and analysis guide for [BOOK/ARTICLE/PLAY/POEM] tailored specifically for [GRADE LEVEL]. The goal is to provide a resource that not only summarizes the text but also unlocks its deeper meaning, themes, and relevance for students, serving as a bridge between reading and full comprehension.

The guide must be structured as follows:

1 **The "Why" it Matters**: A brief, impactful explanation of the value of a strong summary (e.g., "An effective summary provides the foundational scaffolding for literary analysis. It ensures all students have a firm grasp of the narrative arc and key details, freeing up cognitive load to engage with the more complex tasks of interpreting themes, analyzing authorial choices, and connecting the text to the wider world.").

2 **Overarching Learning Objective**: State the primary analytical goal for a student reading this text (e.g., "Students will be able to identify and analyze how the author uses symbolism to develop the central theme of identity," or "Students will be able to trace the protagonist's journey and explain how their motivations change the plot.").

3 **The Summary & Analysis**: This section should be presented in a clear, student-friendly format.

   ◇ **At a Glance**: A one-to-two-sentence "blurb" that captures the essence of the text.

   ◇ **Who & What (The Plot)**: A concise, paragraph-length summary of the main characters and plot, avoiding unnecessary spoilers unless they are central to analysis.

   ◇ **How & Why (The Author's Craft)**: A brief analysis that moves beyond what happens to how and why it matters. Mention 1-2 key literary devices used (e.g., "The author uses a first-person point of view to create a sense of intimacy and subjectivity," or "Foreshadowing in the early chapters builds suspense for the climax.").

4 **Key Concepts & Analysis**:

   ◇ **Central Themes**: A list of 3-5 central themes

(e.g., "The Conflict between Individuality and Conformity," "The Loss of Innocence"). For each theme, provide a one-sentence explanation.

◇ **Character Insights**: Brief notes on 2-3 main characters, focusing on their key motivations, traits, or development.

◇ **Key Takeaways & Relevance**: A list of 3-5 student-centered takeaways that connect the text to the students' lives or broader societal issues (e.g., "This story challenges us to consider: what truly defines a family?" or "The character's decision illustrates the enduring conflict between personal desire and social duty.").

5 **The "Why" Behind This Strategy**: A concluding summary explaining how this multi-layered approach—combining summary with analysis of craft, themes, and relevance—builds critical literacy skills, fosters deeper classroom discussion, and empowers students to formulate their own evidence-based interpretations.

6 **Bonus**: Pro-Tip for Engagement - Discussion Starter: Add a short section with 2-3 compelling, open-ended questions designed to spark critical thinking and classroom dialogue (e.g., "Was the protagonist's final decision justified? Why or why not?" or "How would the story change if it were told from a different character's perspective?").

The final output should be a complete, plug-and-play literary guide that provides a clear summary while simultaneously modeling advanced literary analysis for students, making it an indispensable tool for pre-reading, review, or discussion preparation.

Prompt: Essay Prompt generator

Simple Prompt: "Act as an expert in Writing Pedagogy and Curriculum Design. Your task is to create a comprehensive set of essay prompts for the topic of [TOPIC] tailored to [GRADE LEVEL/COURSE]".

- **Tone**: Academic, Thought-Provoking
- **Format**: Essay Prompt List
- **Platform**: Classroom Assignments, Exam Papers
- **AI Role**: Academic Coach
- **Prompt Goal**: Inspire students to write with depth and clarity
- **Tags**: #essayPrompts, #studentWriting, #criticalThinking
- **Prompt Variables**: {topic/subject}, {grade/level}

Best Practice Tip: Pair essay prompts with a grading rubric so expectations are transparent.

Enhanced Prompt:

Act as an expert in Writing Pedagogy and Curriculum Design. Your task is to create a comprehensive set of essay prompts for the topic of [TOPIC] tailored to [GRADE LEVEL/COURSE]. The goal is to generate prompts that are not only thought-provoking but are also strategically designed to assess specific writing modes, cognitive skills, and learning objectives, providing students with meaningful opportunities to demonstrate their understanding.

The guide must be structured as follows:

1 **The "Why" it Matters**: A brief, impactful explanation of the core principle behind effective prompt design (e.g., "A well-crafted essay prompt is an assessment engine. It simultaneously evaluates a student's mastery of content, their ability to structure a coherent argument, and their skill in employing the conventions of different writing modes. It moves beyond summary to demand critical synthesis and original thought.").

2 **Alignment to Learning Objectives & Standards**: List the 2-3 primary skills or standards this set of prompts is designed to assess (e.g., "CCSS.ELA-LITERACY.W.9-10.1: Write arguments to support claims using valid reasoning," "Analyze the cause-and-effect relationships within a complex system," "Synthesize information from multiple sources to form an original conclusion.").

3 **The Prompts**: A Strategic Mix Provide a list of 5-7 distinct prompts. Ensure there is a deliberate mix of the following types to serve different assessment purposes:

   ◇ **Argumentative/Persuasive**: Requiring a defensible thesis and evidence-based reasoning.

   ◇ **Expository/Analytical**: Requiring explanation, analysis, and synthesis of ideas.

   ◇ **Narrative/Reflective**: Requiring the use of narrative techniques to convey a personal insight or story connected to the topic.

   ◇ **Research-Based/Synthesis**: Requiring the integration of external sources and data.

4 **Breakdown for Each Prompt**: For every prompt, include:

◇ **The Prompt**: Phrased clearly and precisely, so it is student-friendly and unambiguous..

◇ **Writing Mode**: Label it as Argumentative, Expository, Narrative, or Research-Based.

◇ **Cognitive Demand**: Identify the primary cognitive skill required (e.g., Analyze, Evaluate, Compare, Synthesize, Reflect).

◇ **The "Why" it's Important**: A brief explanation of the specific skill or understanding this prompt assesses and its long-term academic value (e.g., "This prompt requires students to evaluate competing perspectives, a skill crucial for informed citizenship and academic research.").

◇ **Suggested Skills Focus**: One specific element of writing the teacher might choose to emphasize during evaluation (e.g., "Thesis clarity," "Use of textual evidence," "Counter-argument development," "Narrative pacing.").

5 **The "Why" Behind This Strategy**: A concluding summary explaining how offering a choice of strategically varied prompts increases student engagement, allows for differentiation by student interest and strength, and provides a more complete picture of the class's overall writing and analytical abilities.

6 **Bonus**: Pro-Tip for Teachers - The Assessment Kit: Add a short section with actionable advice for implementation.

◇ **Providing Choice**: "Offer students a choice between 2-3 prompts to increase ownership and motivation."

◇ **Rubric Building**: "For consistent grading, create a single rubric focused on core skills (e.g., Thesis, Evidence, Organization, Conventions) that can be applied across all prompts, adjusting the weight of the 'Suggested Skills Focus' as needed."

◇ **Scaffolding**: "For learners who need support, provide a basic outline or sentence starters aligned with the chosen prompt."

The final output should be a complete, plug-and-play essay prompt kit that is easy to integrate into a unit plan, provides clear assessment goals, and challenges students to produce their best writing.

Prompt: Sample Research Paper Generator

Simple Prompt: "Act as a expert in Academic Writing Instruction and Curriculum Design. Your task is to create a comprehensive, annotated research paper exemplar on [TOPIC] for [GRADE LEVEL]".

- **Tone**: Academic, Structured
- **Format**: Research Paper Outline + Intro
- **Platform**: Student Projects, Academic Writing
- **AI Role**: Research Mentor
- **Prompt Goal**: Provide models for research paper writing
- **Tags**: #research, #academicWriting, #studentSupport
- **Prompt Variables**: {topic}, {citationStyle}, {grade/level}

Best Practice Tip: Teach students to use the outline as a starting point and expand with their own research.

Enhanced Prompt:

Act as an expert in Academic Writing Instruction and Curriculum Design. Your task is to create a comprehensive, annotated research paper exemplar on [TOPIC] for [GRADE LEVEL]. The goal is to produce a model that is meticulously structured, properly sourced, and, most importantly, explicitly designed to teach students the rhetorical moves, organizational strategies, and citation conventions required for academic success. This should be a "mentor text," not just a finished product.

The guide must be structured as follows:

1. **The "Why" it Matters**: A brief, impactful explanation of the principle behind using mentor texts (e.g., "A strong sample paper is a pedagogical scaffold. It demystifies the writing process by making abstract guidelines concrete. Students learn the 'moves' of academic writing not by being told, but by seeing them effectively executed in context, which builds confidence and provides a replicable template for their own work.").

2. **Overarching Writing Objectives**: List the 3-4 primary writing and research skills this exemplar is designed to teach (e.g., "Skill: Integrating evidence smoothly using signal phrases. Skill: Structuring a logical argument within a paragraph. Skill: Formatting in-text citations and a Works Cited page according to MLA 9th edition guidelines. Skill: Crafting a thesis statement that is specific, arguable, and previews the paper's structure.").

3. **The Annotated Mentor Text**: A Deconstructed Exemplar

    ◇ **Paper Length & Format**: Specify a realistic length (e.g., "3-4 pages") and citation style (e.g., MLA, APA, Chicago).

    ◇ **The Paper**: Write a complete, high-quality sample research paper on the given topic, appropriate for the grade level.

    ◇ **Annotations**: This is the critical element. In the margins or in a separate key, include detailed annotations that highlight and explain the following:

        • **Rhetorical Moves**: (e.g., "Here, the writer uses a 'funnel' introduction to move from a general concept to the specific thesis.")

- **Transitional Phrases**: (e.g., "Note the use of 'Conversely,' to signal a shift to a contrasting viewpoint.")

- **Source Integration**: (e.g., "This is an example of a direct quote introduced with a signal phrase that establishes the author's credibility.")

- **Structural Labels**: (e.g., "Topic Sentence," "Evidence," "Analysis," "Concluding Sentence.").

4 **The "Why" Behind Each Section**: A separate section that deconstructs the purpose and long-term impact of each major part of the paper.

◇ **Introduction & Thesis**: "A strong thesis acts as a roadmap for the entire paper. Mastering this skill is crucial for all persuasive writing, from college essays to professional reports."

◇ **Body Paragraphs (TEAL/TEAC structure)**: "Lear-ning to structure a paragraph around a claim, supported by evidence and analysis, teaches logical thinking and clarity of expression, which are valuable in any field."

◇ **Countcrargument & Rebuttal**: "Addressing opposing views demonstrates critical thinking and intellectual honesty, making the overall argument more robust and credible."

◇ **Conclusion**: "A conclusion that synthesizes points and discusses broader implications shows an ability to think big-picture, a key marker of advanced scholarship."

◇ **Works Cited/References**: "Proper citation is a non-negotiable academic skill that respects intellectual property and allows others to verify your research."

5 **The "Why" Behind This Strategy**: A concluding summary explaining how this "apprenticeship" model—where students analyze a master example—accelerates skill acquisition, reduces anxiety, and produces more confident and competent academic writers than simply providing a rubric alone.

6 **Bonus**: Pro-Tip for Teachers - How to Use This Exemplar: Add a short section with actionable classroom activities.

◇ **"Notice and Note"**: "Have students work in groups to find and label all the elements highlighted in the annotations themselves."

◇ **"I Do, We Do, You Do"**: "Use a section of the paper as a model ('I Do'), then co-write a new paragraph on a different topic as a class ('We Do'), before having students write their own ('You Do')."

◇ **Checklist for Success**: "Provide students with a checklist derived from the annotations (e.g., 'Does my introduction end with a clear thesis?', 'Have I used a signal phrase for every quote?') to use during self-editing."

The final output should be a complete, plug-and-play instructional kit that empowers the teacher to effectively demystify the research paper process and empowers students to become proficient academic writers.

Complete Output (Mini-Exemplar)

Title: The Impact of Social Media on Teen Mental Health
Format: MLA, 3–4 pages (Grade 10 Example)

Introduction (Annotated)

"In today's digital age, social media has become a central part of adolescent life. While platforms like Instagram, TikTok, and Snapchat provide entertainment and connection, they also bring challenges that affect mental health. Recent studies reveal a growing link between heavy social media use and increased anxiety, depression, and low self-esteem among teens. This paper argues that while social media can foster community and self-expression, its overuse negatively impacts teen mental health by fueling social comparison, disrupting sleep, and reducing face-to-face interaction."

Annotations:

- **Funnel Introduction**: Starts broad (digital age → teens → mental health).
- **Thesis Statement**: Arguable, specific, previews structure → three impacts (comparison, sleep, interaction).
- **Signal of Balance**: Acknowledges positive aspects before stating the main claim (critical thinking).

Body Paragraph 1 (Annotated): "One major way social media affects teen mental health is through social comparison. Researchers from the American Psychological Association (APA) found that teens who spend more than three hours a day on social platforms are more likely to report feelings of inadequacy (Smith, 2022). Constant exposure to curated images of peers' achievements, vacations, and appearance creates unrealistic standards. As psychologist Jean Twenge explains, 'Adolescents are comparing their behind-the-scenes with everyone else's highlight reel' (Twenge, 2019, p. 47). This leads to lowered self-esteem and heightened anxiety, as teens may feel they can never measure up."

Annotations:

- **Topic Sentence**: Clear link to thesis ("social comparison").

- **Evidence Integration**: Uses two sources (APA + Twenge) with signal phrases.

- **Quote Framing**: Author name included → credibility.

- **Analysis**: Explains why the evidence matters ("unrealistic standards" → self-esteem/anxiety).

- **Structural Labels**: [Topic Sentence → Evidence → Explanation → Analysis → Conclusion of Paragraph].

Why This Output Matters: Students see a model of academic writing with annotations that explain the "moves." They learn:

- how to build a funnel intro

- how to craft a thesis with 3 points

- how to integrate quotes with signal phrases

- how to analyze instead of just summarizing evidence

Prompt: Research Topic Generator

Simple Prompt: "Act as an expert in Inquiry-Based Learning and Academic Research. Your task is to create a comprehensive menu of compelling research topics for [SUBJECT/COURSE] designed for [GRADE LEVEL]".

- **Tone**: Supportive, Academic
- **Format**: Topic List (x5)
- **Platform**: Student Projects, Research Assignments
- **AI Role**: Topic Advisor
- **Prompt Goal**: Help students choose meaningful and manageable research topics
- **Tags**: #researchTopics, #studentProjects, #learning
- **Prompt Variables**: {subject}, {grade/level}, {focusArea}

Best Practice Tip: Encourage students to narrow broad topics into specific, researchable questions.

Enhanced Prompt:

Act as an expert in Inquiry-Based Learning and Academic Research. Your task is to create a comprehensive menu of compelling research topics for [SUBJECT/COURSE] designed for [GRADE LEVEL]. The goal is to generate topics that are not only engaging but are also strategically designed to be appropriately scoped, academically rigorous, and capable of fostering critical thinking, effective source evaluation, and original synthesis in students.

The guide must be structured as follows:

1 **The "Why" it Matters**: A brief, impactful explanation of the core principle behind effective research (e.g., "A well-designed research topic is a gateway to authentic intellectual inquiry. It should ignite curiosity, challenge students to navigate complex information, and empower them to form and defend their own evidence-based conclusions, mirroring the work of real-world scholars and professionals.").

2 **Overarching Research Skills Objectives**: List the 2-3 primary academic and cognitive skills this research project is designed to develop (e.g., "Skill: Formulate a focused, researchable question. Skill: Evaluate the credibility and bias of primary and secondary sources. Skill: Synthesize information from multiple perspectives to build a coherent argument.").

3 **The Research Menu**: A Spectrum of Inquiry Provide a list of 5-7 distinct research topics. Ensure there is a deliberate mix of the following types to cater to diverse interests and align with different sub-fields within the subject:

   ◇ **Historical Analysis**: Examining past events, figures, or movements through a specific lens.

   ◇ **Scientific Investigation**: Exploring a natural phenomenon, a technological innovation, or an ethical dilemma in science.

   ◇ **Social/Cultural Study**: Analyzing contemporary issues, trends, or systems within society or culture.

   ◇ **Problem-Based/Solution-Oriented**: Researching a real-world problem and proposing a viable solution.

◇ **Comparative Analysis**: Comparing and contrasting two ideas, works, systems, or time periods.

4 **Breakdown for Each Research Topic**: For every topic, include:

◇ **Topic Title**: A broad, engaging area of inquiry (e.g., "The Ethics of Artificial Intelligence").

◇ **Driving Question**: A specific, open-ended, and arguable question that focuses the research (e.g., "To what extent should AI be used in criminal justice sentencing, and what ethical safeguards must be implemented?").

◇ **The "Why" it's Important**: A brief explanation of the topic's relevance and the long-term skills it develops (e.g., "This topic requires students to grapple with emerging technology, apply ethical frameworks, and consider societal impact, fostering critical citizenship and future-ready skills.").

◇ **Starting Points & Key Concepts**: 2-3 suggestions for where to start research or essential ideas to explore (e.g., "Key Concepts: algorithmic bias, recidivism, due process. Start with: research papers from MIT Technology Review's Ethics section.").

5 **The "Why" Behind This Strategy**: A concluding summary explaining how this menu of focused, question-driven topics promotes deeper engagement than a simple report, teaches students how to navigate information overload, and ultimately leads to the production of a more original and insightful final product.

6 **Bonus**: Pro-Tip for Teachers - Scaffolding Success: Add

a short section with actionable advice for setting students up for success.

◇ **The Proposal**: "Require a brief research proposal before beginning. Students must submit their driving question, a preliminary bibliography, and a thesis statement to ensure their topic is focused and feasible."

◇ **Source Bootcamp**: "Dedicate class time to teaching source evaluation. Use the CRAAP (Currency, Relevance, Authority, Accuracy, Purpose) test or a similar framework."

◇ **Check-Ins**: "Implement milestone check-ins for annotated bibliographies, outlines, and drafts to provide formative feedback and prevent last-minute work."

The final output should be a complete, plug-and-play research module that provides a clear path from curiosity to a polished academic product, empowering students to conduct meaningful, self-directed inquiry.

Complete Output:

Sample Research Topics for High School Social Studies

The "Why" it Matters: A strong research topic transforms student work from a generic report into authentic inquiry. With focused driving questions, students learn to think like scholars—evaluating sources, weighing evidence, and drawing conclusions.

The Research Menu: A Spectrum of Inquiry

- **Topic Title**: The Ethics of Artificial Intelligence

    ◇ **Driving Question**: To what extent should AI be used in criminal justice sentencing, and what ethical safeguards must be implemented?

    ◇ **Why It's Important**: Students explore technology's social impact, applying ethical reasoning and critical citizenship.

    ◇ **Starting Points & Key Concepts**: Algorithmic bias, recidivism, due process. Begin with MIT Technology Review articles on AI ethics.

- **Topic Title**: The Impact of Social Media on Teen Mental Health

    ◇ **Driving Question**: Does social media use contribute more to connection or isolation among teenagers?

    ◇ **Why It's Important**: Directly relevant to student lives and SEL (social-emotional learning). Encourages evidence-based discussion of a controversial issue.

    ◇ **Starting Points & Key Concepts**: Cyberbullying, dopamine response, sense of belonging. Start with CDC adolescent well-being studies.

- **Topic Title**: Climate Change and Local Communities

    ◇ **Driving Question**: How is climate change affecting [your city/region], and what local solutions could mitigate its impact?

    ◇ **Why It's Important**: Connects global issues to students' local context, promoting problem-solving and civic engagement.

◇ **Starting Points & Key Concepts**: Rising temperatures, flood/drought patterns, renewable energy. Start with EPA or local government climate reports.

- **Topic Title**: Civil Rights Then and Now

  ◇ **Driving Question**: How do current social justice movements (e.g., Black Lives Matter) compare to the U.S. Civil Rights Movement of the 1960s?

  ◇ **Why It's Important**: Teaches comparative historical analysis and connects past struggles to modern issues of equity and justice.

  ◇ **Starting Points & Key Concepts**: Protest strategies, legislation, media impact. Start with MLK's speeches and contemporary op-eds.

- **Topic Title**: The Role of Music in Cultural Identity

  ◇ **Driving Question**: How does music reflect and shape the identity of a generation or community?

  ◇ **Why It's Important**: Encourages students to examine culture through an accessible lens (music) while practicing academic analysis.

  ◇ **Starting Points & Key Concepts**: Protest songs, hip-hop culture, folk traditions. Start with Smithsonian Folkways archives.

The "Why" Behind This Strategy: This set provides breadth (ethical, cultural, historical, scientific) and depth (driving questions + starting points). It ensures students can connect research to both curriculum goals and personal interests, leading to authentic and high-quality academic work.

Bonus: Pro-Tip for Teachers – Scaffolding Success:

- Require each student to submit their driving question + 3 preliminary sources for approval.
- Use milestone check-ins to guide progress (proposal → annotated bibliography → outline → draft).
- Allow some student choice to increase ownership and motivation.

Prompt: Critical Thinking Exercise

Simple Prompt: act as an expert 'Critical Thinking Architect' and 'Pedagogical Designer'. Your task is to design structured learning experiences that guide students from rote memorization to advanced analysis, evaluation, and creation.

- **Tone**: Analytical, Encouraging
- **Format**: Activity Plan
- **Platform**: Classroom, Workshops
- **AI Role**: Critical Thinking Mentor
- **Prompt Goal**: Develop analytical and evaluative skills in students
- **Tags**: #criticalThinking, #studentGrowth, #classroomActivities
- **Prompt Variables**: {topic}, {grade/level}, {learningGoals}

Best Practice Tip: Ask students to justify their reasoning in groups before sharing with the whole class.

Enhanced Prompt:

You are an expert 'Critical Thinking Architect' and 'Pedagogical Designer'. Your expertise lies in creating structured intellectual experiences that move students beyond memorization into the realms of analysis, evaluation, and creation. You draw from frameworks like Bloom's Taxonomy, Socratic Questioning, and Project Zero's Thinking Routines to design activities that make thinking visible and develop lifelong cognitive skills.

Primary Objective: To design a complete, plug-and-play critical thinking activity that is engaging, easy to implement, and rigorously aligned with the goal of fostering deeper, more informed analysis and decision-making in students.

Core Requirements for the Critical Thinking Challenge: The activity must be a fully-formed blueprint with the following components:

1. The Foundational Principle ("The Why"):

- Articulate the core cognitive skill this activity is designed to develop (e.g., "Evaluating Evidence," "Understanding Perspective," "Identifying Bias," "Constructing Logical Arguments").

- Explain its real-world importance in one compelling sentence. (e.g., "This exercise trains the mind to dissect complex claims, a fundamental skill for navigating an information-saturated world.")

2. Learning Objectives & Success Criteria:

- Define 2-3 clear, measurable learning objectives using Bloom's verbs.

  ◇ **Example**: "By the end of this activity, students will be able to analyze two opposing arguments on a topic, evaluate the strength of the evidence presented, and construct a reasoned conclusion."

- Provide a simple checklist for students to self-assess their thinking process. (e.g., "I can identify the central claim in each argument. I can list the evidence supporting each claim. I can explain which argument is stronger and why.")

3. The Activity Architecture:

- **Activity Title**: Give the activity a clear, engaging name (e.g., "The Argument Tribunal," "Perspective Shift," "The Source Investigator").

- **Time Allocation**: Recommend a realistic time frame (e.g., 45-60 minutes).
- **Format**: Specify the format (e.g., individual, think-pair-share, small group, whole-class debate).
- **Step-by-Step Instructions**: Provide a numbered, easy-to-follow sequence of steps for the teacher to facilitate, from launch to conclusion.

4. The Rule Framework:

- Establish 3-5 simple, non-negotiable rules to create a safe and productive environment for critical discourse.

  ◇ **Example Rules**: "1. Critique the idea, not the person. 2. You must reference evidence to support your claim. 3. Every group member must contribute one probing question."

5. The Question Bank:

- **Provide a tiered set of probing questions to scaffold thinking at different levels**:

  ◇ **Level 1 (Analytical)**: Questions that break down the topic. (e.g., "What are the key components of this argument?")

  ◇ **Level 2 (Evaluative)**: Questions that assess quality and validity. (e.g., "What assumptions is this argument making? How strong is this evidence?")

  ◇ **Level 3 (Synthetic)**: Questions that encourage new ideas. (e.g., "How could these two opposing viewpoints be synthesized into a new, stronger solution?")

6. Differentiation & Accessibility:

- Provide specific modifications to ensure all learners can

access and engage with the activity.

> ◇ **For Support**: Suggest scaffolds (e.g., a graphic organizer, a pre-filled vocabulary bank, sentence starters).

> ◇ **For Extension**: Provide challenge prompts for early finishers or advanced students (e.g., "Defend the position you personally disagree with.").

7. The Pedagogical Rationale:

- Briefly explain the "why" behind the activity's design, referencing educational theory.

> ◇ **Example**: "This uses a 'Structured Academic Controversy' format to reduce emotional debate and foster collaborative sense-making. The tiered questions are designed to scaffold thinking from lower-order to higher-order cognitive processes, as defined by Bloom's Taxonomy."

8. Assessment & Feedback:

- Suggest a simple formative assessment method to gauge student understanding (e.g., an exit ticket question, a shared class document with key takeaways, teacher observation notes).

- Provide an example of specific, actionable feedback a teacher could give to a student based on their performance in the activity.

Final Output Structure: I require the final activity to be presented as a professional, ready-to-use lesson component.

Output Format:

- **Header**: Activity Title, Target Skill, Time, Grade Level Adaptability

- **Sections 1-8**: As detailed above, with clear headings

and bullet points for easy scanning.

- **Bonus**: Pro-Tip for Implementation: Offer one piece of practical advice for the teacher (e.g., "How to randomly assign groups quickly," "A phrase to use if the discussion gets stalled," "How to effectively debrief the activity as a class").

Final Request: Please generate the complete Critical Thinking Challenge Blueprint.

Complete Output

Critical Thinking Challenge Blueprint

Activity Title: The Argument Tribunal

Target Skill: Evaluating Evidence & Constructing Logical Arguments

Time: 50 minutes

Grade Level Adaptability: Upper Elementary – High School

1. The Foundational Principle ("The Why")

This activity develops the skill of evaluating evidence—essential for academic success and real-world decision-making. Students learn to separate strong reasoning from weak claims, building habits of thoughtful inquiry in an information-heavy world.

2. Learning Objectives & Success Criteria

Objectives

- Analyze two opposing arguments on a topic.

- Evaluate the quality and relevance of evidence presented.
- Construct and defend a reasoned conclusion.

Self-Assessment Checklist

√ I identified the central claim in each argument.

√ I listed at least two pieces of evidence for each side.

√ I explained which argument is stronger and why.

3. The Activity Architecture

- **Format**: Small group →√ whole-class discussion.
- **Time Allocation**:
- **Step 1 (10 min)**: Read two short opposing texts on a current issue (teacher provides).
- **Step 2 (15 min)**: In groups, analyze each argument and record evidence.
- **Step 3 (15 min)**: Each group presents their judgment—Which argument is stronger and why?
- **Step 4 (10 min)**: Whole-class synthesis and teacher debrief.

4. The Rule Framework

- Critique ideas, not people.
- Every claim must be backed by evidence.
- Each group member must contribute one probing question.
- Respect speaking turns—listen actively.

5. The Question Bank

- **Level 1 (Analytical)**: What are the key claims in this

argument? What evidence is used?

- **Level 2 (Evaluative)**: Which assumptions are being made? Is this evidence credible?
- **Level 3 (Synthetic)**: How could the best points from each side be combined into a new solution?

6. Differentiation & Accessibility

- **For Support**: Provide a graphic organizer with "Claim → Evidence → Strength." Include sentence starters like: "One strong point is..."
- **For Extension**: Challenge advanced students to defend the position they personally disagree with.

7. The Pedagogical Rationale

This activity is built on Structured Academic Controversy (Johnson & Johnson), which reduces emotional conflict and emphasizes collaborative reasoning. By scaffolding through Bloom's levels—Analyze → Evaluate → Create—students practice high-order thinking in a safe structure.

8. Assessment & Feedback

- **Formative Assessment**: Exit ticket – "Which argument did you find stronger and why?"
- **Teacher Feedback Example**: "You did a great job identifying evidence. Next time, try to explain why one source is more credible than another."

Bonus: Pro-Tip for Implementation

If the discussion stalls, prompt with: "Can you explain that in another way?" or "What evidence supports your view?" These gentle nudges re-center students on analysis without shutting them down.

Prompt: Differentiated Instruction Strategies

Simple Prompt: "Act as an expert in Universal Design for Learning (UDL) and Inclusive Pedagogy. Your task is to create a comprehensive guide to differentiated instruction strategies for educators".

- **Tone**: Inclusive, Practical
- **Format**: Strategy List (x5)
- **Platform**: Classroom, Teacher Planning
- **AI Role**: Inclusive Educator
- **Prompt Goal**: Ensure accessibility and equity in learning
- **Tags**: #differentiatedInstruction, #inclusiveTeaching, #studentSupport
- **Prompt Variables**: {topic}, {grade/level}, {studentNeeds}

Best Practice Tip: Pair differentiation strategies with formative assessments to monitor progress fairly.

Enhance Prompt:

Act as an expert in Universal Design for Learning (UDL) and Inclusive Pedagogy. Your task is to create a comprehensive guide to differentiated instruction strategies for educators. The goal is to provide a toolkit of manageable, high-impact strategies that allow teachers to proactively design instruction that welcomes variability, removes barriers, and provides multiple pathways for all students to reach rigorous learning goals.

The guide must be structured as follows:

1 **The "Why" it Matters**: A brief, impactful explanation of the core philosophy behind differentiation (e.g., "Differentiation is not about creating 25 individual lesson plans; it is about creating a flexible learning environment where all students can access the same core content and rigorous standards through different avenues. It is the acknowledgment that we all learn differently, and it is the teacher's responsibility to remove the barriers that prevent students from demonstrating what they know. It is the essence of equitable, not just equal, teaching.").

2 **Overarching Goal**: State the primary objective of a differentiated classroom (e.g., "To intentionally design instruction that anticipates and plans for learner variability in readiness, interest, and learning profile, ensuring that every student is appropriately challenged and supported every day.").

3 **The Differentiation Toolkit**: A Framework for Access Provide a list of 5-7 distinct strategies. The strategies must be categorized by what you are differentiating:

   ◇ **Content**: What students learn.

   ◇ **Process**: How students make sense of the ideas.

   ◇ **Product**: How students demonstrate their learning.

   ◇ **Learning Environment**: The context in which learning occurs.

4 **Breakdown for Each Differentiation Strategy**: For every strategy, include:

   ◇ **Strategy Name**: A clear, actionable name.

   ◇ **Category**: Label it as Content, Process, Product, or Environment.

◇ **How to Implement It**: A concise, step-by-step explanation of how to execute the strategy in a live classroom setting. Be specific and practical.

◇ **The "Why" it's Important**: A brief explanation of the specific barrier it removes or need it addresses (e.g., "This strategy provides access to complex texts for struggling readers, allowing them to engage with the same critical ideas as their peers without being halted by decoding challenges.").

5 **The "Why" Behind This Strategy**: A concluding summary explaining how this proactive, flexible approach to lesson design is more sustainable and effective than a reactive, "one-size-fits-all" model. It reduces the need for last-minute interventions, increases student engagement and ownership, and creates a truly inclusive classroom community where all learners can thrive.

6 **Bonus**: Pro-Tip for Teachers - Start Small & Build:

◇ **Tier Your Questions, Not Your Lessons**: The simplest way to start is to pre-plan questions for a whole-group discussion at different levels of complexity (Bloom's Taxonomy).

◇ **Use "Must-Do" and "May-Do" Tasks**: This provides structure for some and choice and extension for others, all within the same activity.

◇ **Leverage Peer Power**: Strategic grouping is a form of differentiation. Pair students for complementary strengths.

◇ **Focus on "Low-Prep, High-Impact" Strategies**: The best differentiation is often subtle and doesn't require creating tons of new materials.

The final output should be a complete, plug-and play differentiation toolkit that empowers educators to meet the diverse needs of their learners without burnout.

Complete Output:

The "Why" it Matters: Differentiation is the fundamental belief that fair isn't everyone getting the same thing; fair is everyone getting what they need to be successful. It moves away from teaching to the mythical "middle" and towards designing learning experiences with built-in flexibility. This approach is not a set of extra tasks but a more intentional way of planning that honors the diverse learners in every classroom.

The Differentiation Toolkit: Strategies for an Inclusive Classroom

1. Strategy: Tiered Assignments

- **Category**: Content / Process

- **How to Implement It**: Create 2-3 versions of the same assignment that focus on the same essential skill and concept but vary in complexity, resources, or support. For example:

  - ◇ **Tier 1 (Support)**: A graphic organizer with sentence starters and a word bank.

  - ◇ **Tier 2 (Grade Level)**: The standard assignment.

  - ◇ **Tier 3 (Extension)**: An assignment that requires deeper analysis or application to a novel scenario.

- **Why It's Important**: It ensures all students are working with the same core idea but at a level of challenge that is appropriate for their readiness. This prevents boredom and frustration and keeps everyone in their "zone of proximal development."

2. Strategy: Learning Stations (Centers)

- **Category**: Process / Environment
- **How to Implement It**: Set up different areas in the room where students engage in various activities all related to the same learning goal. Each station can have a different modality (e.g., a video station, a hands-on manipulation station, a reading station, a teacher-led small group station).
- **Why It's Important**: It naturally differentiates by learning style and readiness. It allows the teacher to pull a small, fluid group for targeted instruction (re-teaching or enrichment) while other students work independently. It also incorporates movement and choice.

3. Strategy: Choice Boards

- **Category**: Product
- **How to Implement It**: Provide a grid of activity options for students to choose from to demonstrate their understanding of a concept. The choices should align with different learning preferences (e.g., write a news article, create a comic strip, record a podcast, design a model, compose a song).
- **Why It's Important**: It increases student motivation and ownership by tapping into their interests and strengths. It allows students to express their knowledge in a way that feels natural to them, providing a more accurate picture of their understanding.

4. Strategy: Flexible Grouping

- **Category**: Process / Environment
- **How to Implement It**: Constantly change how stu-

dents are grouped based on the task. Sometimes group by readiness (homogeneously) for targeted support. Sometimes group by interest (heterogeneously) for projects. Sometimes group randomly for think-pair-shares. Avoid static, ability-based tracking.

- **Why It's Important**: It prevents students from being labeled "high" or "low." It allows all students to work with a variety of peers, seeing themselves as both learners and experts in different contexts. It makes grouping dynamic and purposeful.

5. Strategy: Anchor Activities

- **Category**: Process / Environment
- **How to Implement It**: Establish a set of ongoing, meaningful activities that students can work on independently when they finish assigned work early. These should be engaging and curriculum-based, not just busywork (e.g., review games on Quizlet, independent reading, passion projects, practice packets).
- **Why It's Important**: It manages the classroom efficiently by ensuring that early finishers have purposeful work, freeing the teacher to provide additional support to students who need more time. It promotes student autonomy and time management.

6. Strategy: Varied Text & Resource Levels

- **Category**: Content
- **How to Implement It**: Provide resources on the same topic at different reading levels. This could include articles from Newsela (which provides texts at 5+ reading levels), videos, podcasts, or illustrated guides.
- **Why It's Important**: This provides access to complex ideas for all learners, regardless of their reading proficiency. It allows every student to engage with the same

compelling questions and content without the barrier of a text that is too frustrating.

The "Why" Behind This Strategy: This plan works because it focuses on "low-prep, high-impact" strategies that can be woven into existing lessons. By thinking about differentiation through the lenses of Content, Process, Product, and Environment, you can make small, strategic adjustments that have a massive effect on student engagement and achievement. This approach is sustainable, proactive, and empowers the teacher to be a designer of learning, not just a deliverer of information.

Bonus: Pro-Tip for Teachers - Start Small & Build:

- **The One-Strategy Rule**: Don't try to do it all at once. Next week, pick one of these strategies to try with one lesson.

- **Use Data to Group**: Use a quick exit ticket or pre-assessment to inform your flexible groups for the next day. This makes differentiation data-driven.

- **Student Feedback is Key**: Ask students, "What helped you learn this best?" Their answers will give you your best ideas for differentiation.

Prompt: Visualization Roadmap (Student Edition)

Simple Prompt: Act as a 'Cognitive Performance Coach' and 'Visualization Architect'. Your task is to compose a structured guided visualization exercise tailored to support individuals in achieving an [academic/personal goal].

- **Tone**: Inspirational, Supportive
- **Format**: Visualization Script
- **Platform**: Journaling, Classroom Mindset Exercises
- **AI Role**: Visualization Guide
- **Prompt Goal**: Help students mentally rehearse success and growth
- **Tags**: #visualization, #studentMindset, #growth
- **Prompt Variables**: {goal}, {timeframe}, {format}

Best Practice Tip: Encourage students to write or draw what they visualized afterward for reinforcement.

Enhanced Prompt:

You are a 'Cognitive Performance Coach' and 'Visualization Architect'. Your expertise integrates principles from sports psychology, neuroscience, and mindfulness. You understand that the brain's neural pathways can be strengthened through vivid, multi-sensory visualization, effectively making a future success feel familiar and attainable before it physically occurs. This process, known as functional equivalence, primes the mind and body for optimal performance.

Primary Objective:b To craft a precise, guided mental rehearsal exercise that transforms anxiety about a future challenge into confident expectation by creating a robust neural blueprint for success.

My Goal for Visualization: [Insert specific academic/personal goal, e.g., "delivering a flawless and compelling thesis defense," "achieving a high score on the SAT math section," "successfully leading a crucial student council meeting," "completing a complex art portfolio."]

Core Requirements for the Visualization Protocol:

The protocol must be a complete, step-by-step guided script designed to be read slowly and calmly. It must include the following components:

1. The Pre-Visualization Setup (The Framework for Success):

- **Provide clear instructions for the ideal setting**: a quiet space, a comfortable position, and a suggestion to close one's eyes.

- Include a brief physiological sigh breathing exercise (e.g., "Take a double inhale through the nose, followed by a long, slow exhale through the mouth. Repeat twice.") to rapidly reduce stress and transition into a focused state.

2. Acknowledging the "Before" State (The Anchor of Reality):

- Guide the user to briefly and compassionately acknowledge their current feelings of doubt, anxiety, or challenge related to the goal.

- **Purpose**: To validate these feelings without judgment, using them as a contrast point to make the subsequent success feel more powerful. (e.g., "It's natural to feel a sense of nervousness about something this important. Acknowledge that feeling, give it a name, and for now, gently set it aside.").

3. The "In the Moment" Experience (The Vivid Neural Rehearsal):

- This is the core of the protocol. Construct a detailed, first-person, present-tense narrative of the user successfully navigating the challenge.

- Structure it chronologically, walking through the key stages of the event.

- Crucially, focus on the process, not just the outcome. Describe the feelings of competence, the focused mindset, and the effective actions taken. (e.g., "I am walking into the room, feeling calm and prepared. I greet the panel with a genuine smile. As I begin speaking, my voice is clear and confident. I am thoroughly engaging with the material, answering questions with clarity and poise.").

4. Multi-Sensory Integration (The Immersion Engine):

- **Weave specific sensory details throughout the "In the Moment" narrative to make the experience feel real. Provide a note on how to engage each sense:**

    ◊ **Sight**: What does the environment look like? What are the affirming facial expressions of others?

    ◊ **Sound**: What sounds are present? (e.g., the sound of your own confident voice, positive feedback, the quiet focus of a test room).

    ◊ **Touch/Kinesthetic**: What physical sensations are felt? (e.g., the feel of the podium, steady hands, strong posture, a calm heartbeat).

    ◊ **Emotional Sense**: Most importantly, name the

emotions felt during success (e.g., pride, calm, excitement, flow, fulfillment).

5. The "After" State (The Emotional Reward & Identity Reinforcement):

- Describe the immediate aftermath of the success.
- Focus on the feelings of accomplishment, pride, and relief. Paint a picture of the positive consequences and new opportunities that this success unlocks. (e.g., "I feel a profound sense of pride and lightness. I am receiving congratulations, and I feel a new level of confidence in my abilities. This achievement opens a door to what's next.").
- **Reinforce Identity**: Include a statement that connects the success to the student's identity (e.g., "This confirms it: I am the kind of person who can prepare thoroughly and perform under pressure.").

6. The Return & Integration:

- Gently guide the user back to the present moment.
- Reaffirm the connection between the visualized future and the present self. (e.g., "Bring that feeling of confidence and capability back with you into this room, right now. That potential is in you.").

7. The "Next Step" Catalyst (Bridging Mental and Physical Realms):

- Immediately following the visualization, prescribe one micro-action to be taken within the next 24 hours.
- This action must be a tiny, concrete step that directly contributes to the goal, leveraging the momentum of the visualization. (e.g., "Now, open your calendar and schedule one 25-minute block to work on the first slide of your presentation," or "Review the first chapter of your notes for 15 minutes.").

- **Rationale**: This prevents the visualization from being a passive exercise and creates a tangible link between mental rehearsal and real-world progress.

Final Output Structure: I require the final protocol to be presented as a clean, easy-to-follow script.

Output Format:

- **Title**: The Mental Blueprint for: [Your Specific Goal]

- **Introduction**: A brief note on the science of functional equivalence and its purpose.

- **The Guided Script**: The entire protocol written in second-person, present-tense, calming language, formatted for clarity. Instructions for the guide (e.g., "[Pause]", "[Take a breath]") should be in parentheses.

- **Section Headings**: Use headings to denote the different phases (Setup, Before State, In the Moment, etc.).

- **The Catalyst**: The "Next Step" should be clearly boxed or highlighted at the end.

- **Pro-Tip**: Include one piece of advice for frequency (e.g., "For best results, practice this visualization daily for one week leading up to the event.").

Final Request: Please begin by asking me for my specific goal. Then, generate the complete Mental Blueprint for Success Protocol.

Prompt: Emotional Resilience Builder

Simple Prompt: act an 'Emotional Resilience Architect' and 'Cognitive Behavioral Coach'. Your task is to formulate systematic frameworks that empower individuals to recover from adversity and develop robust mental and emotional capacities, enabling them to effectively manage future challenges with enhanced resilience and adaptability.

- **Tone**: Empathetic, Encouraging
- **Format**: Resilience Framework
- **Platform**: Student Journals, Guidance Counseling
- **AI Role**: Resilience Coach
- **Prompt Goal**: Strengthen student resilience and coping skills
- **Tags**: #resilience, #studentSupport, #growth
- **Prompt Variables**: {setbackType}, {studentContext}, {supportSystems}

Best Practice Tip: Pair resilience exercises with reflective journaling to help students track growth.

Enhanced Prompt:

You are an 'Emotional Resilience Architect' and 'Cognitive Behavioral Coach'. Your expertise integrates principles from positive psychology, Cognitive Behavioral Therapy (CBT), Acceptance and Commitment Therapy (ACT), and mindfulness practices. You design structured frameworks that help individuals not just bounce back from adversity, but also build the mental and emotional muscles to navigate future challenges with greater strength and flexibility.

Primary Objective: To create a comprehensive, step-by step, and evidence-based framework that equips students with the practical tools and cognitive skills to process setbacks, regulate emotions, and cultivate enduring resilience.

Core Requirements for the Resilience Framework:

The framework must be a multi-phase, iterative process presented as a clear toolkit. It must include the following components:

1. The Philosophical Foundation ("The Why"):

- Articulate the core principle that resilience is a learnable skill, not an innate trait.

- Explain the concept of "cognitive agility"—the ability to adapt one's thinking and respond effectively to challenges, which is the cornerstone of resilience.

2. The Immediate Triage Protocol (The "In the Moment" Stabilizer):

- Provide a simple, immediate 3-step action plan for managing the initial wave of negative emotions following a setback (e.g., a failed exam, a rejected application, a social conflict).

  ◇ **Step 1**: Acknowledge & Accept. A technique to name and validate the emotion without judgment (e.g., "Name it to tame it": "I am feeling disappointment and frustration right now, and that's a normal reaction.").

  ◇ **Step 2**: Grounding. A quick sensory grounding exercise to prevent emotional overwhelm (e.g., the 5-4-3-2-1 technique: Name 5 things you see, 4 things you feel, 3 things you hear, 2 things you smell, 1 thing you taste).

  ◇ **Step 3**: De-catastrophize. A simple question to

break out of a negative thought spiral (e.g., "On a scale of 1-10, how big will this feel one week from now? One month?").

3. The Cognitive Reframing Module (The "Thought Laboratory"):

- Move beyond a simple "reframe note" to a practical exercise.

- **Provide a "Thought Record" or "Belief Investigation" worksheet template with columns for**:

    ◊ **The Setback**: The objective event.

    ◊ **Initial Thought/Hot Thought**: The automatic negative belief (e.g., "I'm a failure.").

    ◊ **Cognitive Distortion**: Label the thinking error (e.g., overgeneralization, all-or-nothing thinking).

    ◊ **Balanced/Reframed Thought**: A more evidence-based, compassionate perspective (e.g., "I did not succeed this time, but I have succeeded before. This is one event, not my entire identity.").

4. The Strategic Re-engagement Plan (The "Comeback Protocol"):

- A simple, 3-step plan for moving forward, focused on controllable actions.

    ◊ **Step 1**: Micro-Goal Setting. Define one tiny, actionable next step (e.g., "Review the professor's feedback for 15 minutes," "Send one email to ask for clarification").

    ◊ **Step 2**: Resource Identification. Identify what/who could help (e.g., "Office hours, a study group, a tutor, a calming playlist for focus").

◇ **Step 3**: Scheduled Action. Pre-commit to a specific time to execute the micro-goal (e.g., "I will do this at 4 PM today in the library.").

5. The Accountability & Support Ecosystem:

- Provide options for creating an accountability system.

  ◇ **Personal Accountability**: A journaling prompt for weekly review (e.g., "What was one challenge this week and how did I work through it?").

  ◇ **Social Accountability**: A guide for a "Resilience Buddy" check-in. Provide specific questions to ask a friend (e.g., "What's one thing you're struggling with? How can I support you?").

  ◇ **Institutional Support**: Normalize seeking help by listing relevant support resources (e.g., "Academic advising, counseling center, mentorship programs").

6. The "Post-Traumatic Growth" Perspective:

- Reframe the entire process not as just "getting back to normal" but as an opportunity for growth.
- Include guiding questions for a future-focused reflection (e.g., "What did I learn about myself from navigating this? What skill did I strengthen? How might this help me in the future?").

Final Output Structure: I require the final framework to be presented as a practical, visually clear guide that a student could easily implement.

Output Format:

- **Title**: The Emotional Resilience Builder Framework

- **Subtitle**: A Step-by-Step Guide to Navigating Setbacks and Building Strength
- **Visual**: Present the framework as a flow chart or a clear cycle with phases: 1. Triage -> 2. Reframe -> 3. Re-engage -> 4. Reflect & Grow.
- **Detailed Sections**: For each phase, provide the specific tools, scripts, and exercises as outlined above.
- **Worksheets**: Include templates for the "Thought Record" and "Micro-Goal Setting."
- **Key Takeaway**: A one-sentence mantra (e.g., "Resilience isn't about avoiding the fall; it's about learning how to get up with more wisdom and strength.").

Final Request: Please generate the complete Emotional Resilience Builder Framework.

Complete Output

The Emotional Resilience Builder Framework

Subtitle: A Step-by-Step Guide to Navigating Setbacks and Building Strength

1. The Philosophical Foundation ("The Why")

- **Core Principle**: Resilience is not a fixed trait—it is a skill that can be learned, practiced, and strengthened.
- **Cognitive Agility**: Students who learn to adapt their thinking can better navigate challenges, recover faster, and apply these skills across academic and personal contexts.

2. The Immediate Triage Protocol (The "In the Moment" Stabilizer)

- **3-Step Action Plan**:

◇ **Acknowledge & Accept**: Name the emotion to reduce its intensity. "I am feeling overwhelmed right now. That's a normal reaction."

◇ **Grounding Exercise (5-4-3-2-1)**: Name 5 things you see, 4 you feel, 3 you hear, 2 you smell, 1 you taste.

◇ **De-catastrophize**: Ask: "Will this still feel as big in a week? In a month?"

3. The Cognitive Reframing Module (The "Thought Laboratory")

- **Thought Record Template**:

  ◇ **Setback**: What happened?

  ◇ **Hot Thought**: The automatic belief (e.g., "I'll never succeed").

  ◇ **Distortion**: Identify the thinking error (e.g., all-or-nothing thinking).

  ◇ **Reframed Thought**: Balanced, compassionate response (e.g., "This was one challenge, not the whole story of my abilities.").

4. The Strategic Re-Engagement Plan (The "Comeback Protocol")

- **3 Steps to Move Forward**:

  ◇ **Micro-Goal**: Define one small action (e.g., "Review notes for 15 minutes").

  ◇ **Resources**: Identify supports (tutor, peer, playlist, quiet space).

  ◇ **Scheduled Action**: Pre-commit to a time and place.

5. The Accountability & Support Ecosystem

- **Personal Accountability**: Journal prompt — "What challenged me this week, and how did I work through it?"
- **Social Accountability**: Pair with a "Resilience Buddy" to ask: "What's one struggle? How can I support you?"
- **Institutional Support**: Encourage use of counseling services, study groups, and teacher check-ins.

6. The Post-Traumatic Growth Perspective

- **Reflection Questions**:

  ◇ What did I learn about myself?

  ◇ What strengths emerged through this challenge?

  ◇ How might this help me in the future?

- **Identity Reinforcement**: "I am someone who can face setbacks and grow stronger from them."

Visual Flow (Cycle)

Triage → Reframe → Re-Engage → Reflect & Grow

7. Worksheets

Thought Record Table
Event	Hot Thought	Distortion	Reframed Thought
Micro-Goal Plan
Goal	Resource	Time Scheduled

8. The "Why" Behind This Strategy

This framework reframes setbacks as opportunities to practice adaptive thinking. Instead of reacting impulsively, students gain tools to pause, reflect, and act intentionally. This builds long-term coping strategies that extend beyond school into careers and life.

Key Takeaway

"Resilience isn't avoiding the fall—it's learning how to rise each time with greater strength."

Bonus: Pro-Tip for Teachers & Counselors

Use Reflective Journaling as a consistent follow-up. Even 5 minutes of writing can deepen the impact of these practices by making growth visible over time.

Prompt: Growth Trucking Journal

Simple Prompt: Act s a 'Journaling Architect' and 'Cognitive Habit Designer'. Your responsibility is to develop a structured growth journal, featuring daily prompts that facilitate reflection, express gratitude, and monitor progress.

- **Tone**: Reflective, Motivational
- **Format**: Journal Template / Daily Prompts
- **Platform**: Printable Journals, Notion, Student Apps
- **AI Role**: Journaling Mentor
- **Prompt Goal**: Support students in tracking growth and learning progress
- **Tags**: #journaling, #reflection, #studentGrowth
- **Prompt Variables**: {focusAreas}, {journalFormat}, {frequency}

Best Practice Tip: Encourage weekly review of journal entries to help students see their progress.

Enhanced Prompt:

You are a 'Journaling Architect' and 'Cognitive Habit Designer'. Your expertise lies in using structured reflection to enhance metacognition—the ability to think about one's own thinking. You design journaling systems that transform passive experience into active learning, solidifying memory, building self-awareness, and fostering a growth mindset through evidence-based prompts and routines.

Primary Objective: To design a complete, plug-and-play journaling system that moves beyond simple diary entries to become a strategic tool for accelerating student learning, tracking progress, and building resilience.

Core Requirements for the Journaling System: The system must be a multi-layered framework that includes daily, weekly, and monthly components to capture both granular progress and big-picture growth.

1. The Philosophical Foundation ("The Why"):

- **Articulate the core principle**: "Externalization drives internalization." The act of writing down thoughts, learnings, and feelings solidifies neural pathways, making knowledge and self-awareness more durable and accessible.

- Explain how consistent journaling builds metacognitive skills, which are proven to be a key differentiator in academic and personal success.

2. The Daily Prompt Bank (The Core Practice):

- Provide a curated set of 5 daily prompts, categorized by their specific cognitive or emotional purpose. Do not just list prompts; explain their design goal.

  - ◇ **The Learning Laser (Cognitive Reflection)**: A prompt focused on solidifying a key concept learned that day. (e.g., "What was the most important thing you learned today? Explain it in your own words as if teaching it to someone else.")

  - ◇ **The Struggle Spotter (Metacognitive Awareness)**: A prompt that normalizes challenge and identifies learning edges. (e.g., "What was one thing you found difficult or confusing? What strategy did you use to work through it?")

◇ **The Momentum Maker (Progress Tracking)**: A prompt focused on acknowledging effort and micro-wins. (e.g., "What is one small step you took today toward a bigger goal?")

◇ **The Gratitude Anchor (Positive Psychology)**: A prompt to train the brain to spot the positive. (e.g., "What is one thing you are grateful for about your learning experience today? (a person, a resource, a moment of clarity)")

◇ **The Tomorrow Template (Forward Planning)**: A prompt to create intention for the next day. (e.g., "Based on today, what is one specific thing you want to focus on or improve tomorrow?")

3. The Execution Protocol ("The How-To"):

- Provide a clear, ritual-based guide for implementation.

  ◇ **Step 1**: The Cue. Recommend an specific time and trigger for the journaling habit (e.g., "Immediately after your last class of the day," or "As part of your nightly shutdown routine").

  ◇ **Step 2**: The Container. Specify the ideal format (e.g., "A dedicated notebook," "A digital document," "A note-taking app like Notion with a template"). Emphasize consistency over medium.

  ◇ **Step 3**: The Cadence. Set a time limit to prevent overwhelm and ensure sustainability (e.g., "This entire practice should take no more than 5-7 minutes. Use a timer. Focus on brevity and clarity, not essay-writing.").

4. The Synthesis & Review Schedule (The "Check-In"):

- To prevent the journal from becoming a forgotten log, design a mandatory review mechanism.

  - ◊ **Weekly Review (15 minutes)**: Every Sunday, review the week's entries to identify patterns, recurring challenges, and celebrate weekly wins. Provide 2-3 guiding questions for this (e.g., "What was a common theme this week? What is one challenge I overcame?").

  - ◊ **Monthly Audit (30 minutes)**: At the end of the month, perform a higher-level audit. Provide a template for this:

    - ▪ **My Biggest Learning**:

    - ▪ **My Most Effective Strategy**:

    - ▪ **An Area I Want to Grow In Next Month**:

    - ▪ **How I've Surprised Myself**:

5. The Pedagogical Rationale ("The Why Behind the Strategy"):

- Explain the science behind the system's design.

  - ◊ **Example**: "This system is designed using the 'Elaboration' and 'Retrieval Practice' learning techniques. Writing explanations in your own words (Elaboration) and weekly reviews (Retrieval) dramatically increase long-term retention. The gratitude prompt leverages 'Positive Neuroplasticity' to build motivation and resilience."

Final Output Structure: I require the final system to be presented as a ready-to-implement kit.

Output Format:

- **Title**: The Metacognitive Growth Journal System
- **Header**: Core Philosophy, Time Commitment, Required Tools
- **Section 1**: The Daily Prompt Bank (With explanations for each prompt)
- **Section 2**: The Execution Protocol (The step-by-step guide)
- **Section 3**: The Review Templates (Weekly and Monthly review questions formatted in a easy-to-copy way)
- **Pro-Tip**: Offer one piece of advice for success (e.g., "Don't break the chain. Focus on consistency over perfection. Even writing one sentence for each prompt is a victory.").

Final Request: Please generate the complete Metacognitive Growth Journal System.

Complete Output

The Metacognitive Growth Journal System

Core Philosophy: "Externalization drives internalization." Writing down reflections makes thinking visible, strengthens memory, and builds self-awareness.

Time Commitment: 5–7 minutes daily, 15 minutes weekly, 30 minutes monthly.

Required Tools: A notebook, printable template, or digital workspace (e.g., Notion, Google Docs, or a journaling app).

Section 1: The Daily Prompt Bank Each prompt is paired with its purpose to build habits of reflection and growth.

- The Learning Laser (Cognitive Reflection)

    ◇ **Prompt**: "What was the most important thing you learned today? Explain it in your own words as if teaching it to someone else."

    ◇ **Why**: Reinforces understanding and long-term memory through self-explanation.

- The Struggle Spotter (Metacognitive Awareness)

    ◇ **Prompt**: "What was one thing you found difficult or confusing? How did you approach it?"

    ◇ **Why**: Normalizes challenge, builds awareness of strategies, and highlights growth edges.

- The Momentum Maker (Progress Tracking)

    ◇ **Prompt**: "What is one small step you took today toward a bigger goal?"

    ◇ **Why**: Builds motivation by focusing on micro-wins and forward momentum.

- The Gratitude Anchor (Positive Psychology)

    ◇ **Prompt**: "What is one thing you are grateful for about today's learning?"

    ◇ **Why**: Strengthens resilience by training the brain to notice positives.

- The Tomorrow Template (Forward Planning)

    ◇ **Prompt**: "Based on today, what is one specific thing you want to focus on tomorrow?"

    ◇ **Why**: Creates intentional carryover, giving each

day a starting point.

Section 2: The Execution Protocol ("The How-To")

- **Step 1 – The Cue**: Tie journaling to an existing routine, such as right after last class or before bed.
- **Step 2 – The Container**: Use one consistent format (paper journal, digital template, or app).
- **Step 3 – The Cadence**: Limit to 5–7 minutes. The goal is clarity, not essays. Brevity builds consistency.

Section 3: The Review Templates (Check-In System)

- **Weekly Review (15 minutes)**: Guiding Questions:

  ◇ What was a recurring theme or insight this week?

  ◇ What challenge did I work through?

  ◇ What am I most proud of from this week?

- **Monthly Audit (30 minutes)**: Template:

  ◇ **My Biggest Learning**: _____

  ◇ **My Most Effective Strategy**: _____

  ◇ **An Area I Want to Grow In Next Month**: _

  ◇ **How I Surprised Myself**: _____

Section 4: The Pedagogical Rationale

This system uses Elaboration (re-explaining learning in your own words) and Retrieval Practice (weekly/monthly reviews) to boost retention. The Gratitude Anchor strengthens positive neuroplasticity, making students more resilient. Combined, these strategies develop critical metacognitive skills that help students self-regulate and thrive academically and personally.

Pro-Tip for Success

"Don't break the chain." Even if you only write one sentence per prompt, consistency compounds. Small, steady effort builds powerful long-term growth.

Prompt: Health Habit Tracker

Simple Prompt: Act as an Academic Performance Optimizer and Behavioral Habit Scientist. Your task is to design a structured 30-day habit tracking program emphasizing sleep, exercise, and nutrition, which will serve as a foundational approach to promote cognitive enhancement, memory consolidation, and effective emotional regulation.

- **Tone**: Supportive, Practical
- **Format**: Habit Tracker / 30-Day Plan
- **Platform**: Student Wellness Apps, Printable Sheets
- **AI Role**: Wellness Coach
- **Prompt Goal**: Promote healthy habits that support academic success
- **Tags**: #wellness, #habitBuilding, #studentHealth
- **Prompt Variables**: {goal}, {studentLevel}, {trackingMethod}

Best Practice Tip: Link habit tracking to classroom rewards or personal goals for stronger motivation.

Enhaced Prompt:

You are an 'Academic Performance Optimizer' and 'Behavioral Habit Scientist'. Your expertise lies in leveraging lifestyle medicine (sleep, nutrition, movement) as a foundational tool for cognitive enhancement, memory consolidation, and emotional regulation. You design habit systems that are sustainable because they are built on the principle of tiny, consistent gains that compound over time.

Primary Objective: To design a complete, evidence-based 30-day habit sprint specifically tailored to enhance a student's academic performance by optimizing one of the three core pillars of health.

My Chosen Wellness Pillar for the 30-Day Sprint: [Select one: Sleep Hygiene, Exercise & Movement, or Nutrition & Hydration]

Core Requirements for the 30-Day Plan: The plan must be a phased, progressive system that prioritizes consistency and gradual adaptation over intensity.

1. The Foundational Principle ("The Why"):

- Articulate the direct, science-backed link between the chosen pillar and academic success.

  - ◇ **For Sleep**: "Sleep is not downtime; it is memory consolidation time. Deep sleep is when the brain transfers information from the hippocampus (short-term storage) to the cortex (long-term storage), fundamentally solidifying learning."

  - ◇ **For Exercise**: "Movement is cognitive fertilizer. It increases BDNF (Brain-Derived Neurotrophic Factor), a protein that promotes neuroplasticity—the brain's ability to learn, adapt, and grow new neural connections."

  - ◇ **For Nutrition**: "The brain is a high-performance organ that consumes 20% of your calories. Food is information that directly impacts focus, energy levels, and neurotransmitter production for stable mood and motivation."

2. The Phased 30-Day Blueprint (The "Plan"):

- Structure the month into three progressive phases to

prevent overwhelm and build mastery.

◇ **Week 1**: Foundation & Awareness. Focus on tiny, non-negotiable baseline habits and self-observation.

◇ **Week 2-3**: Integration & Expansion. Build upon the foundation, adding slightly more challenging elements.

◇ **Week 4**: Mastery & Automation. Solidify the habits and focus on consistency.

◇ **For each day, provide one specific, actionable micro-habit designed to take 5 minutes or less. (e.g., "Day 1**: Set a phone reminder for your target bedtime." "Day 4: Eat one piece of fruit with your breakfast." "Day 7: Take a 10-minute brisk walk.")

3. The Progress Tracking Triad (The "Tracking"):

- **Provide a simple, 3-part tracking method**:

◇ **The Habit Tracker**: A visual method (e.g., a printable calendar to mark an 'X' for each successful day) to leverage the satisfaction of "not breaking the chain."

◇ **The One-Sentence Journal**: A daily prompt to note the subjective impact of the habit (e.g., "How did my focus/energy/mood feel today compared to yesterday?"). This connects the action to a tangible result.

◇ **The Weekly Win**: A instruction to identify and celebrate one small victory every 7 days (e.g., "I felt more alert in my 9 AM class all week").

4. The Accountability Ecosystem:

- **Suggest a multi-layered approach to account-ability**:

    ◇ **Personal Accountability**: The visual tracker itself.

    ◇ **Social Accountability**: A specific script for a "Wellness Buddy" check-in (e.g., "Text your buddy every Friday with your weekly win and one planned habit for the weekend.").

    ◇ **Environmental Accountability**: One tip to design the environment for success (e.g., "Charge your phone away from your bed," "Pack your gym bag the night before," "Place a full water bottle on your desk each morning.").

5. The Cognitive ROI ("The Game-Changer"):

- Quantify the return on investment in terms of academic performance.

    ◇ **Example**: "This isn't just about health; it's about cognitive ROI. Investing 30 minutes in a walk may feel like 'lost study time,' but the resulting 2-hour window of hyper-focused studying that follows (due to increased blood flow and neurotransmitters) means you net gained 1.5 hours of higher-quality learning."

Final Output Structure: I require the final plan to be presented as a clean, motivating, and ready-to-execute guide.

Output Format:

- **Title**: The 30-Day Academic Wellness Sprint: Optimizing [Chosen Pillar] for Peak Performance

- **Header**: The Science, The Promise, The Time Commitment (<5 mins/day)

- **The Phased Blueprint**: A clear table with columns

for Day, Micro-Habit, and Phase.

- **The Tracking Toolkit**: Templates for the habit tracker and the one-sentence journal prompt.

- **The Accountability Framework**: Clear instructions for each layer of accountability.

- **Final Pep Talk**: A concluding note on the power of consistency and self-compassion.

Final Request: Please begin by asking me to select my wellness pillar (Sleep, Exercise, or Nutrition). Then, generate the complete 30-Day Academic Wellness Sprint plan.

Category 6: School Communication & Administration

Theme

This category emphasizes the importance of clear, inclusive, and effective communication throughout the school community. It is designed to assist teachers, administrators, and school leaders in building strong connections with parents, students, staff, and other stakeholders.

By fostering open dialogue and ensuring that information is shared in a transparent manner, the school ecosystem remains unified and collaborative. The focus is on keeping everyone informed and aligned with school goals and initiatives, thereby promoting trust and a sense of shared purpose.

Category Goal

The objective for this category is to provide schools with a comprehensive set of ready-to-use communication tools. These resources include parent notes, staff memos, community letters, and official announcements.

Each tool is designed to support and enhance trust, ensure transparency, and foster collaboration among all members of the school community. By equipping schools with these communication assets, the goal is to facilitate effective information sharing and strengthen relationships within the entire ecosystem.

Mini-Index: School Communication & Administration

Prompt	AI Role	Prompt Goal
Parent Communication Note	Educational Communication/Family Engagement expert	Write respectful updates for parents
Parent-Teacher Conference Agenda	Family-School Partnership & Collaborative COM. expert	Structure effective parent-teacher meetings
Classroom Newsletter Idea Generator	Educational Marketing specialist	Suggest engaging newsletter ideas
Cross-Cultural Guidelines (School)	Global Education Strategist / Inclusive Systems Architect	Promote inclusivity & sensitivity in communication
Event Speech Writer (School)	Chief Speechwriter / Rhetorical Strategist	Provide inspiring speeches for school events
Staff Communication Memo	School Administrator/Internal COM. Strategist	Draft professional memos for staff updates
Student Progress Report Comment Bank	Master Teacher / Parent Partnership Architect	Provide sample report card comments
School Announcement Draft	Communications Officer	Draft clear announcements for the community
Classroom Policy Handbook Section	Educational Policy Architect	Write handbook policies for schools
Community Engagement Letter	Community Engagement specialist	Strengthen ties between schools & the community

Detailed Prompts

Prompt: Parent Communication Note

Simple Prompt: Act as an expert in Educational Communication and Family Engagement. Your task is to create a comprehensive, ready-to-use template for a parent communication note regarding [CLASSROOM EVENT - e.g., upcoming curriculum unit, class behavior update, academic progress, positive praise].

- **Tone**: Respectful, Supportive
- **Format**: Parent Note
- **Platform**: Email, Letters, Report Cards
- **AI Role**: Teacher Communicator
- **Prompt Goal**: Strengthen communication between teachers and parents
- **Tags**: #parentCommunication, #studentSupport, #education
- **Prompt Variables**: {studentName}, {progress/needs}, {tonePreference}

Best Practice Tip: Keep notes balanced — highlight strengths as well as areas for growth.

Enhance Prompt:

Act as an expert in Educational Communication and Family Engagement. Your task is to create a comprehensive, ready-to-use template for a parent communication note regarding [CLASSROOM EVENT - e.g., upcoming curriculum unit, class behavior update, academic progress, positive praise]. The goal is to craft a message that is transparent, supportive, and strategically designed to foster a collaborative partnership between home and school, all while maintaining a tone of utmost professionalism and respect.

The guide must be structured as follows:

1. **The "Why" it Matters**: A brief, impactful explanation of the core principle behind effective school-to-home communication (e.g., "A strong, proactive parent-teacher partnership is one of the most significant predictors of student success. Clear, respectful communication builds trust, ensures everyone is working from the same information, and aligns efforts to best support the child's growth and well-being. This note is an investment in that critical alliance.").

2. **The Communication Template**: A Structured Framework

   ◊ **Subject Line**: Clear, specific, and neutral (e.g., "Update on [Student Name]' Progress in [Subject]" or "An Invitation to Discuss Our Upcoming [Unit Name] Unit").

   ◊ **Salutation**: Formal and respectful (e.g., "Dear [Parent/Guardian Name],").

   ◊ **Opening of Appreciation & Connection**: A sentence thanking them for their partnership and stating the positive intention of the note.

   ◊ **The Situation (Objective Statement)**: A clear, factual, and non-judgmental description of the event, progress, or observation. Use neutral, observational language.

   ◊ **The Context/Evidence**: Specific, brief examples or data points that illustrate the situation.

◇ **The Shared Goal & Next Steps**: A statement that aligns the teacher and parent on the desired outcome for the student.

◇ **The Call to Action (CTA)**: A specific, direct, and manageable request for the parent.

◇ **Closing of Gratitude & Availability**: A re-affirmation of partnership and an invitation for further dialogue.

◇ **Signature**: Professional sign-off (e.g., "Sincerely," [Your Name], [Subject] Teacher).

3 The **"How-To"**: A Simple 3-Step Personalization Guide

◇ **Step 1**: Insert Specifics: Replace all bracketed placeholders [ ] with the student's name, specific examples, and relevant details.

◇ **Step 2**: Tone Check: Read the note aloud. Does it sound respectful, collaborative, and solution-oriented? Adjust any language that could be perceived as accusatory or defensive.

◇ **Step 3**: Clarity Check: Is the Call to Action crystal clear and easy for a busy parent to accomplish? If not, simplify it.

4 The **Call to Action (CTA)**: Provide a strong, clear, and direct call-to-action.

◇ **Example CTA**: "Please let me know if you have any questions by replying to this email. Would you be available for a brief 10-minute phone call this week to discuss how we can support [Student Name] together? You can schedule a time directly here: [Link to Calendly or other scheduler]."

5 The **"Why" Behind This Strategy**: A concluding

summary explaining how this structured, asset-based approach prevents miscommunication, reduces defensive reactions, and positions the teacher as a caring professional, thereby significantly increasing the likelihood of a positive and productive parental response.

6   **Bonus**: Pro-Tip for Teachers - The 24-Hour Rule & The "Sandwich" Method:

⬦   **The 24-Hour Rule**: For communications about challenging situations, draft the note and then wait 24 hours before sending. This allows you to ensure the message is objective and calm.

⬦   **The "Sandwich" Method (For Difficult Topics)**: Structure the body of the note as: 1. Positive Opener (a genuine strength), 2. The Concern (factual and objective), 3. Hopeful Forward-Looking Closing (reiterating support and partnership).

The final output should be a complete, plug-and-play communication kit that empowers a teacher to initiate difficult or routine conversations with confidence, clarity, and compassion, ultimately strengthening the critical home-school connection.

Complete Output:

Subject Line: An Update on [Student Name]'s Progress in [Subject/Area]

Dear [Parent/Guardian Name],

I'm writing to you today as a partner in [Student Name]'s education. I so appreciate your support and wanted to touch base to ensure we're aligned in helping them have a successful and positive year.

I wanted to share an update regarding [The Specific Situation - e.g., their recent engagement during our math lessons]. I've noticed that [Provide a specific, objective example - e.g., they have been hesitant to participate in group problem-solving and their last two homework assignments were not completed].

My goal for every student, including [Student Name], is to ensure they feel confident and supported in their learning. I am reaching out because I believe that by working together, we can help them overcome this hurdle and rediscover their excitement for the subject.

My call to action for you is simple: Could you please touch base with [Student Name] about this? Perhaps ask them about their feelings toward the current material. Any insight you can provide would be invaluable. Please let me know if you have a moment for a brief follow-up conversation.

Thank you for your partnership. I am committed to supporting [Student Name] and am confident we can help them thrive.

Sincerely,

[Your Name]

[Your Subject] Teacher

[School Name]

[Email Address]

[Phone Number]

Prompt: Parent-Teacher Conference Agenda

Simple Prompt: Act as an expert in Family-School Partnership and Collaborative Communication. Your task is to create a comprehensive Parent-Teacher Conference agenda for a [GRADE LEVEL] teacher meeting about [STUDENT NAME].

- **Tone**: Professional, Structured
- **Format**: Meeting Agenda
- **Platform**: School Events, Parent Meetings
- **AI Role**: School Coordinator
- **Prompt Goal**: Provide structure for effective parent-teacher meetings
- **Tags**: #parentTeacher, #communication, #agenda
- **Prompt Variables**: {studentName}, {grade/level}, {topics}

Best Practice Tip: Share the agenda with parents in advance so they can prepare questions.

Enhanced Prompt:

Act as an expert in Family-School Partnership and Collaborative Communication. Your task is to create a comprehensive Parent-Teacher Conference agenda for a [GRADE LEVEL] teacher meeting about [STUDENT NAME]. The goal is to design a structured conversation that moves beyond simply reporting grades to building a genuine partnership, fostering a shared understanding of the child's strengths and needs, and co-creating a clear, actionable plan for support.

The guide must be structured as follows:

1 **The "Why" it Matters**: A brief, impactful explanation of the core principle behind effective conferences (e.g., "A parent-teacher conference is a strategic partnership meeting. Its goal is to align the two most important teams in a child's life—home and school—around a unified plan for the student's success. A well-structured agenda ensures this short time is productive, positive, and focused on collaborative problem-solving, rather than just delivering a one-way report.").

2 **Overarching Goal**: State the primary objective for this conference (e.g., "To establish a trusting relationship with the family, celebrate the child's unique strengths, honestly discuss areas for growth, and leave with a shared, simple action plan that everyone—teacher, parent, and student—can commit to.").

3 **The Conference Agenda**: A Structured Conversation Flow: The agenda must be designed as a step-by-step guide to a 15-20 minute meeting, with dedicated time for each phase:

   ◇ **Phase 1**: Build Rapport & Set the Tone (The Positive Opener)

   ◇ **Phase 2**: Share Data & Celebrate Strengths (The Evidence)

   ◇ **Phase 3**: Discuss Goals & Areas for Growth (The Collaborative Focus)

   ◇ **Phase 4**: Create a Shared Action Plan (The Forward Momentum)

   ◇ **Phase 5**: Q&A and Next Steps (The Closing)

4 **Breakdown for Each Agenda Item**: For every phase, include:

◇ **Time Allocation**: A suggested time budget to keep the meeting on track.

◇ **Teacher Script/Talking Points**: Suggested language to open each section professionally and empathetically.

◇ **Purpose**: The reason for this part of the conversation.

◇ **Materials to Have Ready**: The specific data points or work samples to reference.

5 **The "How-To"**: A Simple 3-Step Guide to Personalization

◇ **Step 1**: Gather Evidence. Before the conference, collect 2-3 specific work samples that show a strength and an area for growth. Have current data (grades, assessment scores) ready.

◇ **Step 2**: Identify Focus. Choose one primary academic and one primary social-emotional goal to discuss. Avoid overwhelming parents with a list of issues.

◇ **Step 3**: Draft Your Action Plan. Think of one realistic strategy for school and one for home that could support the chosen goals.

6 **The Call to Action (CTA)**: A strong, clear, and direct call-to-action.

◇ **Example CTA**: "Based on our conversation today, our shared goal is to help [Student Name] improve their confidence in math. Our one action step will be: I will check in with them at the start of independent work to ensure they understand the first problem, and we will practice multiplication facts for 5 minutes each night this week. Could we schedule a brief check-in via email in two weeks to see how this is going?"

7 **The "Why" Behind This Strategy**: A concluding summary explaining how this agenda transforms the conference from a potentially defensive encounter into a solution-oriented collaboration. It ensures the parent feels heard, valued, and equipped with clear steps, ultimately creating a stronger support system for the student.

8 **Bonus**: Pro-Tip for Teachers - Managing the Conversation:

◇ **The Parking Lot**: Have a notepad labeled "Parking Lot." If the parent brings up an unrelated or complex issue, say, "That's an important point. Let me add it to our parking lot so we can address it fully later and stay focused on our agenda for today."

◇ **Talk Less, Listen More**: Aim for a 50/50 conversation. Use open-ended questions like "What are you seeing at home?" to elicit their perspective.

◇ **Always End on a Positive Note**: Reiterate one strength of the child and thank the parents for their time and partnership.

The final output should be a complete, plug-and-play conference kit that empowers a teacher to lead a productive, positive, and partnership-building meeting.

Complete Output:

The "Why" it Matters: The parent-teacher conference is the most important face-to-face interaction you have with a family all year. A structured, empathetic approach ensures this brief meeting builds trust rather than defensiveness. It demonstrates your professionalism, your deep knowledge of their child, and your genuine commitment to working with them, ultimately creating a powerful alliance for the student's benefit.

Parent-Teacher Conference Agenda

Student Name: _____

Date: _____ Time: _____

Teacher's Goal for This Meeting: To build a partnership with you to best support your child's growth and happiness at school.

Phase 1: Build Rapport & Set the Tone (3-4 minutes)

- **Teacher Script**: "Thank you so much for taking the time to meet with me today. I really enjoy having [Student Name] in my class. My goal for our time is to share some of my observations about their progress, hear your insights, and work together to create a plan to support them moving forward. How does that sound?"

- **Purpose**: To welcome the family, establish a collaborative tone, and state the meeting's positive purpose.

- **Materials**: A smile, a positive opening comment about the child.

Phase 2: Share Data & Celebrate Strengths (4-5 minutes)

- **Teacher Script**: "I'd like to start by sharing some of [Student Name]'s fantastic strengths. For example, they are a wonderful collaborator during group projects— [give specific example]. Academically, they are excelling in [specific subject/area] and recently did a great job on [show specific work sample]."

- **Purpose**: To begin with positive affirmations, making parents feel proud and setting a constructive tone for the conversation.

- **Materials**: 1-2 strong work samples, a specific anecdote about a social strength.

Phase 3: Discuss Goals & Areas for Growth (5-6 minutes)

- **Teacher Script**: "As we look at their progress overall, an area where I see an opportunity for growth is in [e.g., turning in completed homework, participating in whole-class discussions, math fact fluency]. This is a common goal for this grade level. For example, on this assignment [show work sample], we can see how strengthening this skill would help. What are your thoughts?"

- **Purpose**: To honestly, but kindly, address a growth area, using evidence and framing it as a shared challenge. Crucially, pause and listen to the parent's perspective.

- **Materials**: A work sample that illustrates the growth area, data if applicable (e.g., assessment scores).

Phase 4: Create a Shared Action Plan (3-4 minutes)

- **Teacher Script**: "How can we work together to support this? Here's one thing I can do in the classroom: [e.g., provide a checklist for homework completion]. Is there something you feel you could try at home? [e.g., sign the agenda each night, practice flashcards for 5 minutes]."

- **Purpose**: To co-create a simple, realistic plan with clear responsibilities for both teacher and parent.

- **Materials**: A pen to write down the agreed-upon actions.

Phase 5: Q&A and Next Steps (2-3 minutes)

◇ **Teacher Script**: "What questions do you have for me that we haven't covered? Thank you again for this great conversation. I will follow up with an email summarizing our plan. Let's touch base again in [e.g., a month] to see how things are going."

◇ **Purpose**: To address any remaining concerns, summarize the plan, and set a timeline for "follow-up".

◇ **Materials**: Your "Parking Lot" notepad for off-topic questions.

The "How-To": A Simple 3-Step Guide to Personalization

1 **Fill in the Blanks**: Insert the student's name and your specific examples into the agenda script.

2. **Choose Your Focus**: Select the one most important strength and growth area to discuss. Prepare your work samples for these.

3. **Brainstorm Strategies**: Before the meeting, think of 1-2 strategies you could suggest for both school and home. Be ready to listen to the parent's ideas.

Call to Action (CTA): "Based on our conversation, our shared goal is to help [Student Name] with [Specific Goal]. Our action plan is: I will [Teacher's Action] here at school, and we will [Parent/Student Action] at home. Let's reconnect via email on [Specific Date] to see how it's working."

The "Why" Behind This Strategy: This agenda works because it honors everyone's time and expertise. It values the parent's unique insight into their child and systematically builds a partnership focused on solutions. This structure prevents the meeting from becoming a monologue or a list of problems, ensuring you both leave feeling aligned, empowered, and optimistic about supporting the student.

Bonus: Pro-Tip for Teachers - Managing the Conversation:

- **The Power of "I Notice"**: Use observational language. "I notice they often hesitate to start writing," is less judgmental than "They refuse to do their work."

- **Take Notes**: Jot down the parent's suggestions on the agenda sheet. This shows you value their input.

- **Send a Follow-Up Email**: Within 24 hours, send a brief thank-you email summarizing the action plan. This creates a written record and shows you are accountable.

Prompt: Classroom Newsletter Idea Generator

Simple Prompt: Act as an expert in Educational Marketing and Family Engagement. Your task is to create a comprehensive guide to design a classroom newsletter for [GRADE LEVEL] that transcends a simple information sheet.

- **Tone**: Friendly, Informative
- **Format**: Newsletter Ideas (x5)
- **Platform**: Email, Printouts, School Portals
- **AI Role**: Content Planner
- **Prompt Goal**: Provide engaging newsletter content for parents and students
- **Tags**: #newsletter, #classroomUpdates, #schoolCommunication
- **Prompt Variables**: {classroom}, {frequency}, {audience}

Best Practice Tip: Include student contributions (quotes, drawings) to make newsletters more personal.

Enhanced Prompt:

Act as an expert in Educational Marketing and Family Engagement. Your task is to create a comprehensive guide to design a classroom newsletter for [GRADE LEVEL] that transcends a simple information sheet. The goal is to generate a publication that actively builds a strong, informed, and collaborative community around student learning, making parents feel connected, valued, and equipped to support their child's education.

The guide must be structured as follows:

1. **The "Why" it Matters**: A brief, impactful explanation of the core principle behind a strategic newsletter (e.g., "A classroom newsletter is not a chore; it's your premier marketing tool for building parent partnership. An effective newsletter transforms passive recipients into active allies by demystifying what happens in the classroom, celebrating collective wins, and providing a clear window into their child's world. This proactive communication builds immense trust, reduces misunderstandings, and creates a unified front for supporting student success.").

2. **Overarching Goal**: State the primary objective of this communication (e.g., "To create a consistent, predictable, and joyful touchpoint that informs families about curriculum, celebrates student achievement, and provides actionable tips for supporting learning at home, thereby strengthening the home-school connection and fostering a positive classroom culture.").

3. **The Newsletter Blueprint**: A Recurring Framework Provide a list of 5-7 distinct section ideas. These sections should form a reliable, repeated structure for every edition:

    ◇ **Curriculum & Learning Focus**: Looking forward and backward at academic content.

    ◇ **Celebration & Community**: Highlighting student and class achievements.

    ◇ **Actionable Support**: Giving parents a clear, easy role to play.

    ◇ **Logistics & Reminders**: Essential housekeeping information.

4 **Breakdown for Each Newsletter Section**: For every section idea, include:

- ◇ **Section Title**: A catchy, consistent header name.

- ◇ **Core Content**: A 1-2 sentence description of what belongs in this section.

- ◇ **The "Why" it's Important**: A brief explanation of the long-term impact on parent engagement and the classroom community (e.g., "This section moves parents from asking 'What did you do at school today?' to asking specific, informed questions like 'Tell me about your science experiment with the plants!' This transforms dinner table conversations and deepens student recall.").

- ◇ **Pro-Tip for Execution**: A practical piece of advice for creating this section efficiently and effectively.

5 **The "Why" Behind This Strategy**: A concluding summary explaining how this structured, value-driven approach to communication significantly increases the newsletter's open and read rates, preemptively answers parent questions, and systematically builds a culture of transparency and shared celebration that makes the entire school year run more smoothly.

6 **Bonus**: Pro-Tip for Teachers - The Efficiency Framework:

- ◇ **Create a Template**: Build a reusable template in Google Docs or Canva with these sections pre-labeled. This saves countless hours.

- ◇ **Delegate to Students**: Make the newsletter a class job. Students can write the "What We

Learned" section, draw pictures, or choose the "Student Spotlight."

◇ **Go Digital & Green**: Use a free platform like Smore or Canva to create beautiful, mobile-friendly digital newsletters that track who opens them.

◇ **Consistency Over Perfection**: It's better to send a short, simple newsletter every other week on schedule than to stress over a perfect monthly one that is always late. Set a rhythm and stick to it.

The final output should be a complete, plug-and-play newsletter kit that provides a clear framework for building a powerful and positive home-school partnership.

Complete Output:

The "Why" it Matters: In the busy lives of families, a newsletter is your best chance to control the narrative about your classroom. It's your opportunity to highlight the positive, explain the purpose behind your work, and strategically enlist parents as partners. A great newsletter reduces anxiety, builds excitement for learning, and turns your classroom into a community.

The Newsletter Blueprint: Your Recurring Sections

1. Section: This Week in Learning: A Look Back & Ahead

- **Core Content**: A brief summary of the key concepts covered in the past week (e.g., "We explored fractions as parts of a whole") and a preview of what's coming next (e.g., "Next, we'll be starting our novel study on Charlotte's Web").

- **Why It's Important**: This provides crucial context. It

helps parents connect the dots between isolated home-work assignments and the bigger learning goals, making them feel informed and less surprised by the content their child is engaging with.

- **Pro-Tip**: Write this once in your lesson plan and co-py-paste it into the newsletter. Involve students in writing the "Look Back" summary as a reflection activity.

2. Section: Spotlight On... (Student Work & Achievements)

- **Core Content**: A photo of a stellar student project, a quote from a student's powerful writing excerpt, or a shout-out for demonstrating core values (e.g., perseverance, kindness). Rotate which students are featured each week.

- **Why It's Important**: This is the most-read section. It celebrates effort and achievement, making students feel seen and giving parents a specific, positive reason to open the newsletter. It builds a culture of appreciation.

- **Pro-Tip**: This is a perfect job for a classroom "photographer" or "journalist." Always get permission before sharing any student's work or photo.

3. Section: Ask Me About... (Conversation Starters)

- **Core Content**: Provide 2-3 specific, open-ended questions parents can ask their child to spark a meaningful conversation about school (e.g., "Ask me about the hypothesis for our volcano experiment" or "Ask me about the character we met in read-aloud today.").

- **Why It's Important**: This is a game-changer. It moves parents beyond the dead-end question "How was school?" and gives them a key to unlock detailed, enthusiastic conversations about learning, strengthening their connection to their child's day.

- **Pro-Tip**: Base these questions on the most engaging activity of the week. Students will be excited to talk about it.

4. Section: How to Help at Home (Actionable Tip)

- **Core Content**: One simple, concrete suggestion for supporting current learning (e.g., "When reading together, point out punctuation and ask how it changes the character's voice," or "Practice multiplication facts using a deck of cards—instructions here!").

- **Why It's Important**: It empowers parents who want to help but don't know how. It provides a manageable, curriculum-aligned action that makes them feel like an effective partner in their child's education.

- **Pro-Tip**: Keep it simple and link to resources. This shouldn't feel like assigning homework to parents.

5. Section: Save the Date & Reminders

- **Core Content**: A clean, bulleted list of upcoming deadlines, field trips, holidays, and library days.

- **Why It's Important**: This is the essential utility section that prevents missed permission slips and scheduling conflicts. By making it predictable and easy to scan, you ensure parents can quickly find the critical information they need.

- **Pro-Tip**: Use a consistent format and location in every newsletter so parents know where to look.

6. Section: Volunteer & Supply Needs

- **Core Content**: A specific, low-commitment call for help (e.g., "We need 2 volunteers to read with students Tuesday morning," or "We're running low on glue sticks and tissues. Donations are appreciated!").

- **Why It's Important**: It makes it easy for parents to

contribute to the classroom community in ways that are genuinely helpful to you. Specific, small requests are more likely to be filled than a generic "volunteers needed" plea.

- **Pro-Tip**: Use a Sign-Up Genius link to manage volunteers and donations seamlessly.

The "Why" Behind This Strategy: This plan works because it balances celebration, information, and empowerment. It's not just what you need to tell them; it's what they need to hear to feel connected and capable. This strategic approach transforms the newsletter from an obligation into an anticipated community bulletin that builds trust, fosters partnership, and makes your job easier by proactively addressing questions and needs.

Bonus: Pro-Tip for Teachers - The Efficiency Framework:

◇ **Batch the Work**: Dedicate 20 minutes on Friday afternoon to draft the newsletter for the following week. It's fresh in your mind.

◇ **Use a Digital Tool**: Platforms like Smore or Canva offer beautiful education templates and allow you to schedule sends and see who opened it.

◇ **Student-Led**: For older grades, make the newsletter a rotating student job. It's a fantastic authentic writing and design project.

◇ **The Two-Minute Rule**: A parent should be able to read and digest the entire newsletter in two minutes. Keep text concise and use visuals (photos, icons) to break it up.

Prompt: Cross-Cultural Guidelines (School Edition)

Simple Prompt: Act as a 'Global Education Strategist' and 'Inclusive Systems Architect'. Your task is to develop comprehensive communication guidelines for educational institutions serving diverse student populations, with a focus on promoting cultural inclusivity.

- **Tone**: Inclusive, Respectful
- **Format**: Communication Guidelines
- **Platform**: Policy Documents, Teacher Training
- **AI Role**: Global Education Advisor
- **Prompt Goal**: Support inclusivity and cultural awareness in schools
- **Tags**: #crossCultural, #schoolPolicy, #inclusion
- **Prompt Variables**: {schoolContext}, {studentDemographics}, {culturalFactors}

Best Practice Tip: Translate key communications into multiple languages for families when possible.

Enhanced Prompt:

You are a 'Global Education Strategist' and 'Inclusive Systems Architect'. Your expertise lies in designing communication frameworks that move beyond superficial inclusivity to embed cultural responsiveness into the very fabric of an institution's operations. You understand that effective cross-cultural communication is a critical skill that impacts student belonging, parent engagement, staff cohesion, and ultimately, academic outcomes.

Primary Objective: To draft a comprehensive set of guidelines that transforms how a school communicates internally and externally, ensuring it is respectful, equitable, and effective for all members of its diverse community.

Institutional Context:

Define your school type (e.g., public high school with 40+ spoken," "an international baccalaureate primary school," "a community college with a large first-generation student population," "a university with a significant international student cohort."]

Core Requirements for the Communication Framework: The framework must be a practical, principled guide, not just a list of tips. It must include the following components:

1. The Philosophical Foundation ("The Why"):

- **Articulate the core principle**: "Culturally responsive communication is an active strategy for educational equity." It is not about political correctness; it is about ensuring every student, parent, and staff member feels seen, heard, and valued, which is a prerequisite for learning and collaboration.

- Connect this to the school's mission and the research on how belonging and safety directly impact cognitive function and academic performance.

2. The Core Principles Pillars:

- Establish 3-4 foundational pillars that underpin all guidelines. These should be memorable and actionable.

  ◊ **Pillar 1**: Clarity Over Assumption. Assume nothing about your audience's prior knowledge, context, or language fluency.

  ◊ **Pillar 2**: Respect for Context. Recognize that

communication styles (direct vs. indirect, formal vs. informal) are culturally derived and equally valid.

◇ **Pillar 3**: Intentional Representation. Proactively ensure all communications reflect the diversity of the community.

◇ **Pillar 4**: Active Listening. Communication is a dialogue. Prioritize understanding over being understood.

3. The Practical Guidelines Matrix:

- Organize guidelines into a clear matrix for different contexts and audiences.

  ◇ **A. Verbal & Written Communication**:

  ◇ **Language**: Use plain language; avoid jargon, idioms, and acronyms.

  ◇ **Translation**: Provide key communications in the top 3-5 languages of your community. Use professional translators for official documents.

  ◇ **Tone**: Adopt a warm, formal, and inviting tone by default.

B. Non-Verbal Communication & Events:

  ◇ **Scheduling**: Be mindful of major religious and cultural holidays when planning events.

  ◇ **Logistics**: Ensure events are accessible (physically, financially, and socially). Provide clear information on dress codes, food, and expectations.

  ◇ **Symbolism**: Audit visual materials (website, brochures) to ensure diverse representation in imagery and examples.

C. Digital Communication:

    ◇ **Platforms**: Use multiple platforms (email, SMS, app notifications) to reach families with varying tech access.

    ◇ **Format**: Use headers, bullet points, and clear subject lines to enhance scannability.

D. Interpersonal & Conflict Communication:

    ◇ **Approach**: Assume positive intent and seek to understand before being understood.

    ◇ **Protocol**: Provide a clear, safe, and accessible pathway for parents and students to voice concerns or misunderstandings.

4. The Inclusivity in Action Protocol:

- Provide a checklist for auditing and creating inclusive content.

The AUDIT Tool:

    ◇ A - Assume nothing. Have we provided all necessary context?

    ◇ U - Understand the audience. Have we considered their potential perspectives?

    ◇ D - Diversity check. Does this material reflect our diverse community?

    ◇ I - Invite feedback. Is it clear how someone can ask questions or get help?

    ◇ T - Translate key points. Are the most critical elements accessible to all?

5. The Implementation & Training Plan:

- This must be a living document. Provide a strategy for rollout and sustainability.

  ◊ **Launch**: Introduce the framework at a staff meeting, focusing on the "why."

  ◊ **Training**: Offer voluntary workshops on specific skills (e.g., "Writing for Translation," "Hosting Inclusive Events").

  ◊ **Resources**: Create a shared drive with email templates, a cultural calendar, and translation resources.

  ◊ **Champions**: Identify and train a group of "Inclusivity Champions" in each department to model and advise on the guidelines.

6. The "Next-Level" Strategy: Crisis & Change Management:

- Explain how this framework is a strategic asset during major change or crisis (e.g., a market shift, a public incident, a pandemic).
- **Guideline**: "In times of crisis, revert to the core principles with heightened discipline. Clarity prevents panic, Respect builds trust, Representation ensures no one is overlooked, and Active Listening is how you navigate uncertainty. This framework is your blueprint for maintaining community cohesion when it is most fragile."

Final Output Structure: I require the final framework to be presented as a professional, ready-to-implement policy guide.

Output Format:

- **Title**: Framework for Culturally Responsive Communication: [School Name]
- **Header**: Vision Statement, Core Principles, Date of Im-

plementation

- **Section 1**: Our Why (The Philosophical Foundation)
- **Section 2**: Our Pillars (The Core Principles)
- **Section 3**: Our Guidelines (The Practical Matrix, presented in clear tables for each context)
- **Section 4**: The AUDIT Tool (The checklist for creating content)
- **Section 5**: Implementation & Sustainability (The roll-out plan)
- **Appendix**: Links to resources, translation services, and a cultural calendar.

Final Request:

Please begin by asking me for the specific institutional context. Then, generate the complete Framework for Culturally Responsive Communication.

Prompt: Event Speech Writer (School Events)

Simple Prompt: Act as a You are a 'Chief Speechwriter' and 'Rhetorical Strategist' specializing in educational settings. Your task is to Draft a 5-minute speech for [school event/assembly] that is engaging, clear, and audience-appropriate.

- **Tone**: Inspirational, Audience-Centered
- **Format**: Speech Script (5 minutes)
- **Platform**: Assemblies, Graduations, School Events
- **AI Role**: Speechwriter
- **Prompt Goal**: Provide inspiring speeches for school occasions
- **Tags**: #speechwriting, #schoolEvents, #communication
- **Prompt Variables**: {event}, {audience}, {theme}

Best Practice Tip: Include personal anecdotes or student highlights to make speeches more memorable.

Enhance Prompt:

You are a 'Chief Speechwriter' and 'Rhetorical Strategist' specializing in educational settings. Your expertise lies in applying classical rhetorical principles (Ethos, Pathos, Logos) to craft messages that inspire, unite, and move a school community. You understand that a great speech is a strategic communication tool designed to achieve a specific emotional and behavioral outcome.

Primary Objective: To generate a complete, rhetorically

sound 5-minute speech blueprint tailored to a specific school event, ensuring the message is engaging, appropriate for the audience, and drives a desired feeling or action.

Speech Parameters:

1. **Event Type**: [e.g., "Freshman Orientation," "Academic Awards Night," "Veterans Day Assembly," "Pep Rally," "Teacher Retirement Celebration"]

2. **Primary Audience**: [e.g., "Grade 9 students and parents," "Faculty and staff," "The entire student body"]

3. **Core Message/Theme**: [e.g., "Embrace new beginnings," "Celebrate perseverance, not just achievement," "The power of community," "Gratitude for service"]

4. **Speaker's Role**: [e.g., "Principal," "Student Council President," "Honored Teacher"]

Core Requirements for the Speech Blueprint: The blueprint must be a masterclass in speech construction, broken down into the following components:

1. The Strategic Foundation ("The Why"):

- Articulate the primary objective of the speech. Is it to inspire, to welcome, to celebrate, to challenge, or to unite?

- Define the desired emotional takeaway for the audience (e.g., "I want the freshmen to feel excited and confident, not anxious.").

2. Audience Analysis & Tone Setting:

- Provide a brief analysis of the audience's likely mindset and needs for this event.

- Prescribe the appropriate tone (e.g., inspirational, celebratory, solemn, energetic) and pace (words per minute) for a 5-minute speech (~650-750 words).

3. The Structural Architecture:

- Break the speech down into a timed structure.

  ◇ **0:00 - 0:45** | The Hook (The Attention An-
  chor): Provide a compelling opening. Options:
  a provocative question, a relatable short story, a
  surprising statistic, or a powerful quote. It must
  immediately connect to the core theme.

  ◇ **0:45 - 3:30** | The Body (The Argument/Sto-
  ry): Develop the core message through a balanced use of:

    ▪ **Pathos (Emotional Appeal)**: A relat-
    able anecdote or a shared experience that
    creates connection.

    ▪ **Logos (Logical Appeal)**: A clear, log-
    ical progression of 2-3 main points that
    support the theme.

    ▪ **Ethos (Credible Appeal)**: Weave in
    the speaker's own relevant and authentic
    experience to build trust.

  ◇ **3:30 - 4:45** | The Climax & Call to Action (The
  Memorable Takeaway): synthesize the ideas
  into one powerful, memorable sentence or idea.
  The CTA should be clear, actionable, and aspi-
  rational.

  ◇ **4:45 - 5:00** | The Closing (The Resonant Fin-
  ish): Provide a strong, closing statement that
  circles back to the hook or theme, leaving the
  audience with a final, resonant thought.

4. The Rhetorical Devices Toolkit:

- **Integrate specific rhetorical devices to enhance impact**:

    ◇ **Rule of Three**: Listing items in trios for rhythm and memorability.

    ◇ **Anaphora**: Repeating a word or phrase at the beginning of successive clauses for emphasis.

    ◇ **Vivid Imagery**: Using descriptive language to paint a mental picture.

5. Delivery Notes & Pro-Tips:

- Offer 2-3 specific delivery tips for the speaker (e.g., "Pause after the hook for effect," "Make eye contact with different sections of the audience during the main points," "Practice slowing down your pace during the climax.").
- Include one piece of advice for personalization (e.g., "Replace the anecdote with one from your own experience to make it more authentic.").

Final Output Structure: I require the final output to be presented as a professional speechwriter's blueprint.

Output Format:

- **Title**: Rhetorical Blueprint for: [Event Name]
- **Header**: Target Audience, Core Message, Desired Tone, Word Count
- **Section 1**: Strategic Foundation
- **Section 2**: Timed Structural Breakdown (A table with Time Stamp, Section, Purpose, and Sample Text)
- **Section 3**: Rhetorical Analysis (Notes on where and how Ethos, Pathos, and Logos are used)
- **Section 4**: Delivery Protocol (The pro-tips for delivery)

- **Full Speech Draft**: A complete, formatted draft of the speech, written in a natural, spoken-word style, with notations for pauses [pause] and emphasis (emphasis).

Final Request: Please begin by asking me for the Event Type, Audience, Core Message, and Speaker's Role. Then, generate the complete Rhetorical Blueprint for a School Speech.

Complete Output

Rhetorical Blueprint for: Graduation Ceremony

Target Audience: Graduating seniors, paents, teachers, and community members

Core Message: "Celebrate perseverance, not just achievement" Desired

Tone: Inspirational, Warm, Reflective

Word Count: ~700 (5 minutes spoken)

Section 1: Strategic Foundation

- **Objective**: To inspire graduates to see their journey not as a single milestone but as the result of perseverance and resilience.

- **Desired Emotional Takeaway**: Graduates should feel proud, hopeful, and ready to face the unknown with courage. Parents and teachers should feel validated in their support role.

Section 2: Timed Structural Breakdown

Time	Section	Purpose	Sample Text
0:00 – 0:45	Hook (Attention Anchor)	Capture attention, set theme	"What makes today extraordinary isn't just the cap and gown, or the diploma in your hand. It's the journey—the nights you stayed up finishing an essay, the times you failed and tried again, the quiet persistence that no one applauded, but which got you here."
0:45 – 3:30	Body (Main Message)	Develop theme with Pathos, Logos, Ethos	Pathos (Emotion): "Think of your classmates who encouraged you, your teachers who challenged you, your families who believed in you." Logos (Logic): "Research shows perseverance—grit—is the greatest predictor of success, more than talent or IQ." Ethos (Credibility): "As your principal, I have seen you grow—not only in knowledge but in resilience, in empathy, in character."
3:30 – 4:45	Climax & Call to Action	Deliver takeaway + actionable message	"Today, don't just celebrate your diploma. Celebrate your perseverance. Carry it forward. When life tests you—and it will—remember: You've already proven you can endure and rise."
4:45 – 5:00	Closing (Resonant Finish)	Circle back, leave lasting thought	"Graduates, as you step into your future, remember this: Perseverance is your superpower. Wear it as proudly as you wear that cap and gown today."

Section 3: Rhetorical Analysis

- **Ethos**: Speaker builds credibility by referencing years of observation of the class.

- **Pathos**: Personal anecdotes + recognition of families and teachers create emotional resonance.

- **Logos**: Referencing research on perseverance gives logical weight.

- **Rhetorical Devices**:

  - ◇ **Rule of Three**: "In resilience, in empathy, in character."

  - ◇ **Anaphora**: "Celebrate not just the achievement... Celebrate the perseverance."

  - ◇ **Imagery**: "Nights you stayed up finishing an essay," "the quiet persistence no one applauded."

Section 4: Delivery Protocol

- **Pro-Tips**:

  5. Pause after the Hook ("What makes today extraordinary...") to let the weight sink in.

  6. **Vary eye contact**: left, right, center of audience during main points.

  7. Slow down at the Climax to emphasize the call-to-action.

- **Personalization Advice**: Insert a short, true anecdote about a class-wide challenge (e.g., remote learning year, a school project, a championship) to deepen authenticity.

Full Speech Draft (~700 words)

Note: Written in spoken-word style, with [pauses] and (emphasis) markers.

Good evening, graduates, families, and friends.

[Pause]

What makes today extraordinary isn't just the cap and gown you're wearing, or the diploma you'll soon hold in your hands. (It's the journey.) The late nights finishing an essay. The disappointment of a test score that didn't reflect your effort. The quiet persistence that no one applauded, but which got you here today.

[Pause]

Graduates, this moment is a celebration of perseverance.

Think back—every one of you has a story. A challenge you faced, a moment you thought you might not make it through. Maybe it was balancing school and work. Maybe it was speaking up in class for the first time. Maybe it was pushing through during a global pandemic. Whatever your story, the common thread is this: you kept going.

And that's why you are here.

[Pause]

As your teachers and families, we could not be prouder. We've watched you grow—not only in knowledge but in resilience, in empathy, in character. And research tells us something powerful: perseverance—what psychologists call grit—is the greatest predictor of success. More than talent. More than IQ. More than luck.

You have proven that truth already.

[Pause, shift tone to inspirational]

So today, don't just celebrate your diploma. Celebrate your perseverance. Carry it with you. Because life will test you again—

through challenges we cannot yet see. And when it does, remember this day. Remember the nights you pushed through. Remember the failures you turned into fuel. Remember that you've already proven you can endure, and rise.

[Pause]

Graduates—your future is unwritten. But one thing is certain: perseverance is your superpower. Wear it as proudly as you wear that cap and gown today.

Congratulations, Class of [Year]. We believe in you. We are proud of you. And we cannot wait to see the incredible futures you will create.

[Final pause, smile, applause cue]

Prompt: Staff Communication Memo

Simple Prompt: Act as a 'School Administrator' and 'Internal Communications Strategist'. Your task is to draft a clear, empathetic memo template and plan for staff about [school topic, schedule change, policy, or meeting], prioritizing clarity, compliance, and positive reception.

- **Tone**: Professional, Clear
- **Format**: Staff Memo
- **Platform**: Internal Communication, School Admin Docs
- **AI Role**: School Administrator
- **Prompt Goal**: Ensure clear and timely communication with staff
- **Tags**: #staffCommunication, #schoolAdmin, #memos
- **Prompt Variables**: {topic}, {urgency}, {audience}

Best Practice Tip: Keep memos concise — no more than 1 page or 300 words.

Enhanced Prompt:

You are a 'School Administrator' and 'Internal Communications Strategist'. Your expertise lies in crafting messages that are not only clear and professional but also build trust, reduce ambiguity, and drive alignment among faculty and staff. You understand that in a high-stakes environment like a school, effective communication is a core operational function, not an administrative afterthought.

Primary Objective: To generate a complete, empathetic, and action-oriented memo template and strategy for a specific school-related topic, ensuring maximum clarity, compliance, and positive staff reception.

Memo Parameters:

1. **Memo Topic**: [e.g., "Implementation of a new cell phone policy," "Changes to the Wednesday early-release schedule," "Reminder for mandatory professional development," "Update on safety protocols"]

2. **Sender's Name & Title**: [e.g., "Dr. Jane Doe, Principal"]

3. **Level of Urgency/Importance**: [e.g., "High - Action Required," "Medium - For Review and Feedback," "Low - For Your Information"]

Core Requirements for the Communication Framework: The output must be a masterclass in internal communications, broken down into the following components:

1. The Philosophical Foundation ("The Why"):

- **Articulate the core principle**: "Staff memos are trust-building instruments." Every communication either strengthens or erodes staff confidence in leadership. Clarity demonstrates respect for their time and intelligence, while transparency builds buy-in, even for unpopular decisions.

2. The STAKS Memo Template (A Detailed Framework):

- **Provide a template structured around the STAKS protocol for effective messages**:

    ◇ **S - Subject Line**: Clear, specific, and includes the required action or topic (e.g., "ACTION REQUIRED: Review Updated Cell Phone Policy

Draft by Friday").

◇ **T - Top-Line**: The first sentence must state the who, what, and why in 15 words or less. (e.g., "This memo outlines the new cell phone policy, effective next semester, designed to minimize distraction and maximize learning.")

◇ **A - Action & Details**:

- **The Change/Policy**: What is happening? Use bullet points for scannability.

- **The Rationale (The "Why")**: The educational or operational reasoning behind the decision. This is non-negotiable for building understanding.

- **The Timeline**: When does this happen?

- **The Support**: What resources or training are available?

◇ **K - Key Contacts**: Who should questions be directed to? (e.g., "Please direct all questions to Vice Principal Smith.").

◇ **S - Signature**: Standard professional closing.

3. The Empathy & Tone Integration Guide:

- Provide specific language cues to ensure the memo feels empathetic and collaborative, not just top-down.

  ◇ **Acknowledge Impact**: Include a sentence that recognizes the change's effect on staff (e.g., "We understand this represents a shift in practice and appreciate your flexibility.").

  ◇ **Use "We" and "Our"**: Frame the message around shared goals (e.g., "This change supports our collective goal of..."). Avoid accusatory "you" statements.

◇ **Offer a Feedback Channel (if appropriate)**: For non-urgent policies, provide a way for staff to give input.

4. The Personalization Protocol ("The How-To"):

- **A simple 3-step guide to customizing the template**:

  1 **Insert Specifics**: Fill in the bracketed [ ] information in the STAKS template with your exact details.

  2. **Tone Check**: Read the memo aloud. Does it sound like a collaborative leader or a distant administrator? Adjust language for empathy.

  3. **Clarity Check**: Ask a colleague to read it. Can they state the what, why, and action back to you in 10 seconds? If not, simplify.

5. The Unmistakable Call to Action (CTA):

- The CTA must be bolded and placed in a separate paragraph for visibility.
- It should be specific, time-bound, and simple.

  ◇ **Weak CTA**: "Please be aware of this change."

  ◇ **Strong CTA**: "Please read the attached policy draft and provide feedback via this Google Form by Friday, October 27th."

6. The Strategic Rationale ("The Why Behind the Strategy"):

- **Explain how this framework operates as a strategic tool**:

⬦ **Efficiency**: It saves time and prevents mis-communication, reducing the flood of clarifying emails.

⬦ **Leadership Capital**: Transparent communi-cation builds staff trust, which is the currency of effective leadership. It demonstrates competence and respect.

⬦ **Operational Alignment**: It ensures everyone receives the same message at the same time, moving the entire organization in the same di-rection.

Final Output Structure: I require the final output to be pre-sented as a ready-to-use professional toolkit.

Output Format:

- **Title**: The Strategic Staff Communication Framework
- **Header**: Core Principle, STAKS Protocol Explained
- **Section 1**: The Complete Memo Template (Formatted with placeholders [ ] for easy customization)
- **Section 2**: The Empathy Integration Guide (With ex-amples of what to say)
- **Section 3**: The Personalization & Proofing Proto-col (The 3-step checklist)
- **Pro-Tips**:
- **Distribution**: Send memos at a consistent day/time (e.g., Tuesday mornings).
- **Format**: Use PDF for final policies to prevent editing of official documents.
- **Follow-through**: If you ask for feedback, you must acknowledge it and share how it was used.

Final Request: Please begin by asking me for the Memo Topic,

Sender's Name/Title, and Level of Urgency. Then, generate the complete Strategic Staff Communication Framework.

Complete Output

The Strategic Staff Communication Framework

Core Principle: Staff memos are trust-building instruments. Clear, transparent, and empathetic communication demonstrates respect for staff time and strengthens confidence in leadership.

STAKS Protocol Explained:

- **S – Subject Line**: Clear, specific, action-oriented
- **T – Top-Line**: The who, what, and why in 15 words or less
- **A – Action & Details**: The policy/change, rationale, timeline, and supports, presented in bullet points
- **K – Key Contacts**: The point person for questions or follow-up
- **S – Signature**: Professional closing with name and title

Section 1: Complete Memo Template

Subject Line: [ACTION REQUIRED/UPDATE]: [Specific Topic]

Top-Line: This memo outlines [policy/schedule change/topic] effective [date], to support [goal/outcome].

Action & Details:

- **The Change/Policy**: [Brief description of what is changing/being implemented]
- **Rationale**: [Explain the educational/operational reasoning. Keep it student-centered when possible.]

- **Timeline**: [Effective date, deadlines, or phases]
- **Support**: [What training, resources, or assistance will be provided]

Acknowledgment: We recognize this adjustment may impact your daily routines and appreciate your flexibility and professionalism as we implement this.

Key Contact: For questions, please reach out to [Name, Title, Email/Phone].

Call to Action (bolded):

Please [read/review/complete action] by [date/time].

Closing & Signature: Thank you for your continued commitment to our students and our school community.

Sincerely,

[Your Name]

[Your Title]

Section 2: Empathy Integration Guide

Use these phrases to ensure tone remains collaborative:

- **Acknowledge impact**: "We know this represents a change in practice and appreciate your adaptability."
- **Use shared language**: "Together, we can ensure this supports our collective goal of student success."
- **Feedback channel**: "If you have suggestions or concerns, please share them by [method] by [date]."

Avoid:

- "You must comply..." → Replace with → "To ensure smooth implementation, please complete this step by [date]."

Section 3: Personalization & Proofing Protocol

- **Insert Specifics**: Replace placeholders with the memo's exact details.

- **Tone Check**: Read aloud — does it sound professional and collaborative? Adjust as needed.

- **Clarity Check**: Have a colleague skim. Can they restate the what, why, and action in 10 seconds? If not, simplify.

Pro-Tips for Leaders

- **Distribution**: Send memos at a consistent time (e.g., every Tuesday morning). Predictability increases readership.

- **Format**: Use PDF attachments for finalized policies to avoid edits.

- **Follow-through**: If feedback is requested, acknowledge it and explain how it informed the final decision.

- **Keep it concise**: Aim for one page, no more than 300 words.

Prompt: Student Progress Report Comment Bank

Simple Prompt: Act as a You are a 'Master Teacher' & 'Parent Partnership Architect'. Your task is to develop 20 sample comments suitable for student progress reports across various subjects, ensuring each comment appropriately addresses both the students' strengths and areas requiring improvement.

- **Tone**: Constructive, Encouraging
- **Format**: Comment Bank (20 Examples)
- **Platform**: Report Cards, LMS
- **AI Role**: Teacher
- **Prompt Goal**: Provide ready-to-use language for report cards
- **Tags**: #progressReports, #feedback, #studentGrowth
- **Prompt Variables**: {subject}, {studentPerformance}, {tonePreference}

Best Practice Tip: Always include one actionable suggestion alongside praise.

Enhanced Prompt:

You are a 'Master Teacher' and 'Parent Partnership Architect'. Your expertise lies in crafting narrative feedback that is specific, actionable, and empathetic. You understand that progress report comments are a primary tool for building a collaborative relationship with families, focusing on the whole child's growth rather than just a judgment of performance.

Primary Objective: To create a robust bank of narrative comments and a strategic framework that allows teachers to efficiently provide personalized, meaningful feedback that celebrates strengths, outlines a clear path for growth, and invites parental partnership.

Core Requirements for the Comment Bank Toolkit:

The toolkit must be a comprehensive resource that moves beyond generic phrases to offer nuanced, subject-aware, and developmentally appropriate comments.

1. The Philosophical Foundation ("The Why"):

- **Articulate the core principle**: "The goal of feedback is to illuminate the path forward, not just to audit the past." Effective comments translate grades into narratives, transforming a progress report from a verdict into a conversation starter that aligns teacher and parent support for the student.

2. The Comment Bank Architecture:

- Organize the 20+ comments into a clear, navigable structure.

- **Categorize by Focus Area**:

    ◇ **Academic Mastery**: Comments on understanding core concepts, skill application, and quality of work.

    ◇ **Cognitive & Learning Behaviors**: Comments on critical thinking, problem-solving strategies, curiosity, and intellectual risk-taking.

    ◇ **Executive Functioning**: Comments on organization, time management, work habits, and preparedness.

    ◇ **Social-Emotional Learning**: Comments on

collaboration, communication, resilience, and classroom citizenship.

- **Use the "Glow and Grow" Framework**: For each category, provide:

  ◇ **"Glow" Comments (Strengths)**: 2-3 specific, celebratory comments.

  ◇ **"Grow" Comments (Areas for Development)**: 2-3 constructive, forward-looking comments that suggest a clear strategy for improvement.

- Ensure comments are subject-agnostic or can be easily adapted for Math, ELA, Science, etc.

3. The Personalization Protocol ("The How-To"):

- **Provide a simple, 3-step method for teachers to transform a template into personalized feedback**:

  1 **Select & Combine**: Choose one "Glow" and one "Grow" comment that most accurately reflect the student.

  2. **Insert Specific Evidence**: Add one concrete example from class to illustrate each point. (e.g., "...as demonstrated by his insightful questions during our discussion on the novel." or "...I've noticed this when long-term projects are broken into smaller steps.").

  3. **Adjust the Tone**: Ensure the language is consistent with the student's overall performance and the family's communication style (e.g., more direct for serious concerns, more celebratory for high achievers).

4. The Strategic Call to Action (The Partnership Invitation):

- Every comment should end with a specific, positive, and collaborative invitation for parental support.
- Move beyond generic "please help at home" to actionable suggestions.

  ◇ **Example**: "To support [Student's] growth in organization, you could ask her to show you her planner each night this week. Please feel free to reach out if you'd like more specific strategies."

  ◇ **Example**: "I am excited about [Student's] progress in reading. To foster this further, encourage him to tell you about the main conflict in whatever he's reading for fun. Let's continue this partnership."

5. The Pedagogical Rationale ("The Why Behind the Strategy"):

- Explain the research-based reasoning behind this approach.

  ◇ **Efficiency**: A well-organized bank saves teachers countless hours while producing higher-quality, more consistent feedback.

  ◇ **Efficacy**: Specific, strategy-focused comments give students and parents a clear understanding of how to improve, which is more effective than simply identifying what is wrong.

  ◇ **Partnership**: Framing the CTA as an invitation positions the teacher and parent as a team, reducing defensiveness and increasing engagement.

Final Output Structure: I require the final toolkit to be presented as a professional, ready-to-use resource.

Output Format:

- **Title**: The Constructive Comment Bank Toolkit: For Building Powerful Parent Partnerships
- **Header**: Core Philosophy, How to Use This Guide
- **Section 1**: The Comment Bank (Organized in a clear table with columns: Category, Type (Glow/Grow), Sample Comment)
- **Section 2**: The Personalization Protocol (The 3-step guide with examples)
- **Section 3**: Call to Action Phrases (A list of 5-7 adaptable partnership invitations)
- **Pro-Tip**: Advice on maintaining a balance of positivity and constructive feedback, even for students who are struggling significantly.

Final Request: Please generate the complete Constructive Comment Bank Toolkit.

Complete Output

The Constructive Comment Bank Toolkit: For Building Powerful Parent Partnerships

Core Philosophy: The goal of progress report feedback is not simply to document achievement, but to guide growth. Comments should provide clarity, affirm student effort, and invite parents into a supportive partnership.

How to Use This Guide:

- **Select & Combine**: Choose one "Glow" (strength) and one "Grow" (area for improvement).
- **Add Evidence**: Personalize with a classroom example or data point.

- **End with a CTA**: Use one of the partnership phrases to engage parents.

Section 1: Comment Bank (Glow & Grow by Category)

Academic Mastery

Glow

- "[Student] demonstrates strong understanding of core concepts in [subject], applying them accurately and with confidence."
- "[Student] consistently produces high-quality work, showing attention to detail and depth of understanding."
- "[Student] excels at transferring learning to new contexts, especially during [specific project/assessment]."

Grow

- "[Student] is developing their skills in [specific skill/subject] and would benefit from additional practice with [strategy/tool]."
- "[Student] sometimes struggles with retaining key concepts over time; reviewing notes or practicing daily could strengthen this area."
- "[Student] is encouraged to ask more clarifying questions when content feels unclear to ensure deeper understanding."

Cognitive & Learning Behaviors

Glow

- "[Student] demonstrates curiosity by asking thoughtful questions that extend class discussions."
- "[Student] shows persistence in problem-solving, often trying multiple approaches before arriving at an answer."

- "[Student] engages critically with content, offering insightful connections to real-world issues."

Grow

- "[Student] can strengthen their critical thinking by supporting answers with clear evidence."
- "[Student] is encouraged to slow down and reflect before finalizing work to improve accuracy."
- "[Student] could take greater academic risks by sharing tentative ideas during class discussions."

Executive Functioning

Glow

- "[Student] arrives prepared for class, consistently bringing materials and completing assignments on time."
- "[Student] demonstrates strong organizational skills, using planners and tools effectively to manage tasks."
- "[Student] is proactive in managing deadlines, often completing assignments ahead of schedule."

Grow

- "[Student] is working on building stronger time-management skills; breaking larger tasks into smaller steps may help."
- "[Student] sometimes needs reminders to bring necessary materials, but shows improvement when using checklists."
- "[Student] could benefit from establishing a consistent homework routine to reinforce classroom learning."

Social-Emotional Learning

Glow

- "[Student] is a kind and respectful classmate who contributes positively to group work."

- "[Student] demonstrates resilience by bouncing back quickly from setbacks and maintaining a positive outlook."

- "[Student] actively listens to peers, showing empathy and strengthening our classroom community."

Grow

- "[Student] is encouraged to build confidence in sharing ideas during group work and class discussions."

- "[Student] is developing strategies to manage frustration when tasks feel difficult."

- "[Student] can strengthen collaboration skills by balancing listening with contributing their own perspective."

Section 2: The Personalization Protocol

- **Select & Combine**: Choose one Glow + one Grow comment.

- **Insert Evidence**: Add a specific classroom reference:

    ◇ **Example**: "...as demonstrated by her detailed analysis during the science lab write-up."

    ◇ **Example**: "...I've noticed this particularly during long-term projects that require multiple steps."

- **Adjust Tone**: Soften for encouragement, sharpen for clarity depending on context.

Section 3: Call to Action Phrases

- "To support [Student's] growth in [skill], you might try [simple home strategy]. Let's check in again soon."

- "Encouraging [Student] to review their planner nightly could strengthen consistency. Please let me know if you'd like additional strategies."

- "One way to build confidence is to have [Student] explain what they learned at dinner. This reinforces class discussions."

- "To support ongoing growth, a routine of 10 minutes of [skill practice] each evening would be very beneficial."

- "I'd love to hear your insights from home about [Student's] learning habits so we can align strategies."

Section 4: Pro-Tips for Teachers

- **Balance**: Always start with a Glow before introducing a Grow.

- **Specificity**: Avoid generic words like "good" or "bad" — ground comments in skills and behaviors.

- **Consistency**: Use the same structure across students for fairness, but personalize with evidence.

- **Positive Framing**: Phrase challenges as opportunities for growth, not shortcomings.

Prompt: School Announcement Draft

Simple Prompt: Act as a 'Director of School Communications' and 'Crisis & Community Engagement Specialist'. Your task is to draft a school-wide announcement for [event/change] that is clear and engaging.

- **Tone**: Informative, Engaging
- **Format**: Announcement Draft
- **Platform**: Website, PA System, Newsletters
- **AI Role**: Communications Officer
- **Prompt Goal**: Ensure clear messaging for the entire school community
- **Tags**: #announcements, #schoolNews, #communication
- **Prompt Variables**: {event/change}, {audience}, {channel}

Best Practice Tip: Keep announcements short — focus on the who, what, when, where, and why.

Enhanced Prompt:

You are a 'Director of School Communications' and 'Crisis & Community Engagement Specialist'. Your expertise lies in crafting messages that cut through the noise of daily life to ensure critical information is received, understood, and acted upon by a diverse community of students, parents, staff, and caregivers. You understand that a school announcement is a strategic tool for building trust, ensuring safety, and fostering a informed community.

Primary Objective: To generate a complete, clear, and engaging announcement template and strategy for a specific school-related event or change, ensuring 100% clarity and maximum compliance across the entire community.

Announcement Parameters:

1. **Announcement Topic**: [e.g., "Upcoming Parent-Teacher Conferences," "New Drop-Off Procedure," "School Closure Due to Inclement Weather," "Annual Art Show Invitation"]

2. **Sender's Name & Title**: [e.g., "Maria Garcia, Principal"]

3. **Level of Urgency/Importance**: [e.g., "Urgent - Action Required," "Essential Information," "Community Event"]

4. **Target Audience**: [e.g., "K-12 Families & Staff," "Grades 6-8 Parents," "All Faculty"]

Core Requirements for the Announcement Framework: The output must be a masterclass in public communication, broken down into the following components:

1. The Philosophical Foundation ("The Why"):

- **Articulate the core principle**: "Every school-wide announcement is a test of institutional clarity and trust." In a potential information vacuum, a poorly crafted message breeds confusion and anxiety, while a clear, transparent, and timely message builds confidence in school leadership and ensures community safety and alignment.

2. The STAK Announcement Template (A Detailed Framework):

- **Provide a template structured around the STAK protocol for public messaging**:

◇ **S - Subject Line**: Must be incredibly specific and include the topic and required action. (e.g., "ACTION NEEDED: Sign Up for Spring Parent-Teacher Conferences by Friday" or "UPDATE: Revised Drop-Off Procedure Starting Monday, Oct 21").

◇ **T - Top-Line**: The first sentence must state the most critical information: the who, what, when, and why in 20 words or less. (e.g., "This is an important update on our new drop-off procedure, designed to improve student safety, beginning this coming Monday.").

◇ **A - Action & Details**:

  ▪ **The Core Information**: Bulleted or numbered list of key facts. Answer all anticipated questions (Who, What, When, Where, Why, How).

  ▪ **The Rationale (The "Why")**: Briefly explain the reason for the change or event. This builds understanding and reduces resistance.

  ▪ **The Timeline**: Clearly state effective dates, deadlines, or event times.

◇ **K - Key Contacts & Resources**: Who should questions be directed to? Include links to websites, forms, or maps. (e.g., "For questions, view the new traffic flow map on our website or contact Mr. Davis at transportation@school.org.").

3. The Multi-Platform Distribution Strategy:

- **A message is only effective if it is received. Provide a checklist for distribution across the community's preferred channels**:

⬦ **Primary Channel (Email)**: For detailed information.

⬦ **Secondary Channel (SMS/App Alert)**: For urgent messages or a quick reminder with a link to the full email. (e.g., "Weather Alert: Schools closed tomorrow. See email for details.").

⬦ **Tertiary Channels (Verbal/Physical)**: Reinforce the message through morning announcements, social media, and physical flyers.

4. The Personalization Protocol ("The How-To"):

- **A simple 3-step guide to customizing the template**:

  ⬦ **Fill the Blanks**: Insert all specific details into the STAK template's [placeholders].

  ⬦ **The "Grandparent Test"**: Read the announcement aloud. Could someone with no context (like a grandparent helping with childcare) understand exactly what to do? Simplify any confusing jargon.

  ⬦ **Channel Check**: Ensure the message is tailored for each platform (e.g., shorter for SMS, more visual for social media).

5. The Unmistakable Call to Action (CTA):

- The CTA must be in a separate, bolded paragraph.
- It should be specific, time-bound, and simple.

  ⬦ **Weak CTA**: "Please be aware of this change."

  ⬦ **Strong CTA**: "To ensure a smooth transition, please review the new drop-off map on our website by this Friday and discuss it with your student."

6. The Strategic Rationale ("The Why Behind the Strategy"):

- **Explain how this framework operates as a critical operational tool:**

  ◇ **Crisis Prevention**: Clear communication prevents logistical nightmares (e.g., traffic jams from a new procedure) and manages parental anxiety during emergencies.

  ◇ **Operational Efficiency**: A single, well-crafted message reduces the administrative burden of answering hundreds of individual questions.

  ◇ **Community Trust**: Transparency and consistency in messaging demonstrate competent, caring leadership, which is the foundation of a strong school community.

Final Output Structure: I require the final output to be presented as a ready-to-use professional toolkit.

Output Format:

- **Title**: The School-Wide Announcement Framework
- **Header**: Core Principle, STAK Protocol Explained, Distribution Strategy
- **Section 1**: The Complete Announcement Template (Formatted with placeholders [ ] for easy customization)
- **Section 2**: The Multi-Platform Distribution Checklist
- **Section 3**: The Personalization & Clarity Protocol (The 3-step guide including the "Grandparent Test")
- **Pro-Tips**:
- **For Urgent Messages**: Use a recognizable, consistent subject line format for all emergencies (e.g., always start-

ing with "ALERT:").

- **For Events**: Include a link to add the event directly to a digital calendar.

- **Accessibility**: Use a large, legible font and ensure high color contrast.

Final Request: Please begin by asking me for the Announcement Topic, Sender's Name/Title, Level of Urgency, and Target Audience. Then, generate the complete School-Wide Announcement Framework.

Complete Output

The School-Wide Announcement Framework

Core Principle: Every announcement is a test of institutional clarity and trust. Clear, timely communication builds confidence, prevents confusion, and strengthens community engagement.

The STAK Protocol (Explained)

- **S – Subject Line**: Be specific and action-oriented.

- **T – Top-Line**: State the who, what, when, why in ≤20 words.

- **A – Action & Details**: Provide bulleted, scannable facts + rationale.

- **K – Key Contacts & Resources**: Direct readers to the right person/place.

Section 1: Complete Announcement Template

Subject Line: [ACTION REQUIRED/UPDATE/INVITATION]: [Topic/Event Name]

Top-Line: [One sentence summarizing the essential info: who, what, when, where, why.]

Announcement Body:

- **What's Happening**: [Clear description of the event/ change]
- **When**: [Date & Time / Effective Starting Date]
- **Where**: [Location or applicable group]
- **Why**: [Brief rationale—e.g., for safety, efficiency, celebration, community engagement]
- **How**: [Step-by-step or link to instructions/forms if needed]

Call to Action (CTA): Please [specific action, e.g., sign up / attend / review] by [deadline].

Key Contacts: For questions, contact [Name, Title] at [email/ phone]. Additional details available at [link].

Signature:

Sincerely

[Name]

[Title]

Section 2: Multi-Platform Distribution Checklist

- **Primary Channel (Email/Letter)**: Full details, formatted clearly.
- **Secondary Channel (SMS/App Alert)**: Concise reminder, link to full announcement.
- **Tertiary Channels (Verbal/Visual)**:
  - ◇ Morning announcements / PA system.
  - ◇ Social media posts (graphic + key info).

◇ Flyers/posters for hallways or backpacks.

Section 3: Personalization & Clarity Protocol

Step 1: Fill the Blanks – Replace placholders with exact event details.
Step 2: Grandparent Test – Read aloud: could someone with no school context (like a grandparent helping with childcare) understand what to do?

Step 3: Channel Check – Adjust the message for platform:

- Email → Detailed
- SMS → ≤160 characters
- Social Media → Visual + link

Pro-Tips for Administrators

- **Urgent Messages**: Use a consistent "ALERT:" subject line for emergencies.
- **Event Announcements**: Add a link to sync directly into Google/Outlook calendars.
- **Accessibility**: Provide translations in top community languages; use high-contrast text for visibility.
- **Consistency Over Length**: Short, clear, timely beats long and late.

Prompt: Classroom Policy Handbook Section

Simple Prompt: Act as an 'Educational Policy Architect' and 'Classroom Culture Designer'. Write a handbook section on [topic: attendance, homework, classroom behavior]in a clear and professional manner.

- **Tone**: Formal, Clear
- **Format**: Policy Section
- **Platform**: School Handbooks, Teacher Guides
- **AI Role**: Policy Writer
- **Prompt Goal**: Provide standardized guidelines for classrooms
- **Tags**: #policy, #schoolHandbook, #administration
- **Prompt Variables**: {topic}, {grade/level}, {policyRequirements}

Best Practice Tip: Use plain, simple language so both students and parents can easily understand.

Enhanced Prompt:

You are an 'Educational Policy Architect' and 'Classroom Culture Designer'. Your expertise lies in crafting guidelines that are not just rules, but foundational elements for building a predictable, fair, and positive learning environment. You understand that effective policies are teaching tools that promote student agency, self-regulation, and collective responsibility.

Primary Objective: To generate a complete, clear, and pedagogically sound policy section for a classroom handbook that standardizes expectations while fostering a culture of respect, accountability, and academic growth.

Policy Parameters:

1. **Policy Topic**: [Select one: Attendance & Punctuality, Homework & Practice, Classroom Behavior & Community norms, Technology Use, Assessment & Grading]

2. **Grade Level**: [e.g., Elementary, Middle School, High School]

3. **Educational Philosophy**: [e.g., "Restorative Practices," "Growth Mindset," "Student-Centered Learning"]

Core Requirements for the Policy Framework: The policy must be a comprehensive guide that justifies, explains, and operationalizes the rule.

1. The Philosophical Foundation ("The Why"):

- Articulate the core educational principle behind the policy. Move beyond "classroom management" to deeper values.

  ◇ **Example for Homework**: "Independent practice is designed to reinforce learning, build executive functioning skills, and provide data for the teacher. This policy is built on the principle of quality over quantity and mastery over completion."

  ◇ **Example for Behavior**: "Our classroom is a community of learners. This policy is founded on respect for self, others, and our shared space, and utilizes restorative practices to repair harm and rebuild trust."

2. The Policy Statement (The "What"):

- Draft the policy itself in clear, student-friendly language. Use positive framing where possible (e.g., "We will be prepared for class" instead of "Don't forget your pencil").

- **Structure it with clear headings for scannability**:

  ◇ **Our Policy**: A one-sentence, overarching statement.

  ◇ **Student Responsibilities**: A bulleted list of specific, observable student actions (e.g., "Submit homework by 9:00 AM at the designated bin," "Charge your device nightly," "Use kind and inclusive language").

  ◇ **Teacher Responsibilities**: A bulleted list of what the teacher will do to uphold their end of the agreement (e.g., "Provide clear grading rubrics," "Return graded work within one week," "Facilitate restorative circles when needed"). This builds mutual respect and accountability.

3. The Rationale & Impact ("The Why It's Important"):

- Explain the direct link between the policy and student outcomes. Connect the dots for students and parents.

  ◇ **Example for Attendance**: "Consistent attendance is directly correlated with academic achievement. Being on time minimizes disruptions and shows respect for everyone's learning time. The habits of punctuality and reliability are critical life skills for future success in college and career."

4. The Procedures & Protocols (The "How"):

- This is the most critical and often missing section. Detail the step-by-step process for following the policy and addressing violations.

◇ **For Homework**: What is the process for late work? Are there revisions? How are extensions requested?

◇ **For Behavior**: What is the graduated response to disruptive behavior? (e.g., 1. Non-verbal cue, 2. Private conversation, 3. Reflection sheet, 4. Parent contact, 5. Office referral). This removes ambiguity and ensures fair application.

5. The Pedagogical Rationale ("The Why Behind the Strategy"):

- Explain the research and theory informing the policy's design.

  ◇ **Example**: "This policy utilizes a graded response model instead of zero-tolerance. This is because research in educational psychology shows that consequences are most effective when they are logical, related to the infraction, and aimed at teaching the appropriate behavior, not just punishing the negative one. This approach reduces power struggles and preserves the student-teacher relationship."

Final Output Structure: I require the final policy to be presented as a ready-to-implement handbook section.

Output Format:

- **Title**: Classroom Policy: [Policy Topic]
- **Header**: Grade Level, Core Philosophy
- **Section 1**: Our Commitment (The Philosophical Foundation)
- **Section 2**: The Policy (The "What" - formatted with clear sub-headings: Student Responsibilities, Teacher Responsibilities)

- **Section 3**: Why This Matters (The Rationale & Impact)
- **Section 4**: How It Works (The Procedures & Protocols, presented as a clear flow chart or numbered list)
- **Pro-Tip for Implementation**: Advice on how to roll out the policy (e.g., "Co-create the 'Behavior' policy with students on the first day," "Send the 'Homework' policy home for parents and students to sign, acknowledging understanding.").

Final Request: Please begin by asking me for the Policy Topic, Grade Level, and Educational Philosophy. Then, generate the complete Classroom Policy Framework.

Complete Output

Classroom Policy: Homework & Practice

Grade Level: Middle School

Core Philosophy: Growth Mindset & Student-Centered Learning

Section 1: Our Commitment (The Philosophical Foundation)

Homework is not about busywork—it is about building confidence, independence, and mastery. This policy is grounded in the belief that practice outside of class helps students strengthen skills, deepen understanding, and develop lifelong habits of responsibility. We approach homework with a growth mindset, viewing mistakes as opportunities to learn and improvement as the ultimate goal.

Section 2: The Policy (The "What")

Our Policy: Homework is meaningful practice designed to reinforce classroom learning, encourage independent thinking, and prepare students for future lessons.

Student Responsibilities:

- Complete all assigned homework by the due date.
- Show effort and neatness in every assignment.
- Ask questions in class if concepts are unclear.
- Bring completed work to class ready for discussion or review.

Teacher Responsibilities:

- Assign homework that is purposeful, relevant, and reasonable in length.
- Provide clear instructions and due dates.
- Offer timely feedback on submitted assignments.
- Be available for support and clarification.

Section 3: Why This Matters (The Rationale & Impact)

Consistent homework practice builds study skills, accountability, and self-discipline—essential skills for success in higher grades, college, and beyond. Research shows that meaningful, appropriately assigned homework can improve student performance by reinforcing classroom content, encouraging independent problem-solving, and promoting better long-term retention.

Section 4: How It Works (The Procedures & Protocols)

1. **Submission**: Homework is due at the beginning of class on the assigned date.

2. **Late Work**:

   - Accepted within 2 days for partial credit (70%).

- Beyond 2 days, work may not receive credit but should still be completed for practice.

3. **Extensions**: Students may request an extension before the due date for valid reasons (illness, family responsibilities).

4. **Revisions**: Major assignments may be revised once after feedback for improved credit.

5. **Support Options**: Students may attend homework help sessions after school or during advisory if they need additional assistance.

Pro-Tip for Implementation

Introduce this policy in the first week of school and co-create a student-friendly version with your class (e.g., a "Homework Promise" chart signed by students and displayed in the room). Send a parent-friendly summary home and invite questions, ensuring alignment between school and family expectations.

Prompt: Community Engagement Letter

Simple Prompt: "Act as a 'Director of Community Engagement' and 'Partnership Architect' for a school. generate a complete, persuasive, and warm letter template and strategy that successfully invites community members to participate in [school initiative/event]".

- **Tone**: Warm, Respectful
- **Format**: Invitation Letter
- **Platform**: Email, Print, Community Outreach
- **AI Role**: School Leader
- **Prompt Goal**: Strengthen ties between schools and the community
- **Tags**: #community, #schoolEngagement, #partnerships
- **Prompt Variables**: {initiative/event}, {audience}, {benefits}

Best Practice Tip: Highlight mutual benefits — how the community gains as well as the school.

Enhanced Prompt:

You are a 'Director of Community Engagement' and 'Partnership Architect' for a school. Your expertise lies in crafting outreach that moves beyond a simple transaction (e.g., "come to our event") to foster genuine, reciprocal relationships. You understand that sustainable community partnerships are built on mutual benefit, shared values, and clear, respectful communication.

Primary Objective: To generate a complete, persuasive, and warm letter template and strategy that successfully invites community members to participate in a school initiative, framing it as an opportunity for shared investment and mutual benefit.

Partnership Parameters:

1. **School Initiative/Event**: [e.g., "Career Day," "School Garden Volunteer Program," "Literacy Mentorship Initiative," "Spring Festival"]

2. **Target Audience**: [e.g., "Local business owners," "Senior center residents," "Alumni," "Families new to the district"]

3. **Sender's Name & Title**: [e.g., "Michael Chen, Principal of Maplewood Elementary"]

4. **Desired Community Role**: [e.g., "Gues speaker," "Volunteer," "Sponsor," "Participant"]

Core Requirements for the Partnership Framework: The output must be a masterclass in stakeholder engagement, broken down into the following components:

1. The Philosophical Foundation ("The Why"):

- **Articulate the core principle**: "Schools are not islands; they are the hearts of their communities." This initiative is designed to break down the traditional barrier between the school and the street it sits on, creating a powerful feedback loop where community expertise enriches student learning, and student energy revitalizes community spirit. This is about reciprocal value, not a one-sided ask.

2. The A.I.D.A. Letter Template (A Persuasive Framework):

- Provide a template structured around the AIDA copywriting model (Attention, Interest, Desire, Action) to maximize engagement.

◇ **A - Attention (Subject Line & Salutation)**: A compelling subject line and a personalized salutation that establishes a local connection.

- **Subject**: "Invitation to Shape Future Leaders: Be a Mentor at Oak High's Career Day"

- **Salutation**: "Dear Neighbor & Valued Community Partner,"

◇ **I - Interest (The Hook & Shared Value)**: Open with a statement of shared purpose or a relatable observation about the community. Connect the school's mission to broader community well-being.

◇ **D - Desire (The Invitation & Benefits)**:

- **The Ask**: Clearly describe the initiative and the specific, manageable role for the community member.

- **The "What's In It For We"**: Articulate the mutual benefits. What does the student gain? (e.g., real-world perspective). What does the community member gain? (e.g., talent pipeline, fulfilling volunteerism, positive PR, a stronger community).

- **Logistics**: Clearly state the date, time, time commitment, and location.

◇ **A - Action (The Clear CTA)**: A direct, simple call to action with a single, prioritized point of contact.

3. The Personalization Protocol ("The How-To"):

- **A simple 3-step guide to customizing the tem-**

plate for maximum impact:

1. **Hyper-Localize**: Replace generic terms with specific local references. Mention the recipient's business or organization by name in the body of the letter if possible.

2. **Align the Benefit**: Tailor the "What's In It For We" section to the specific audience (e.g., for a tech company, highlight building a future talent pipeline; for a senior, highlight sharing wisdom and legacy).

3. **Authentic Voice**: Read the letter aloud. Does it sound like it's from a real person in your community? Adjust formality to match your school's culture.

4. The Multi-Channel Distribution Strategy:

- **A letter may not be enough. Provide a strategy for reinforcement**:

  ◇ **Primary Channel**: A personalized letter or email.

  ◇ **Secondary Channel**: A shared social media post for wider reach.

  ◇ **Tertiary Channel**: A personal phone call from a PTA member or student to key stakeholders.

5. The Unmistakable Call to Action (CTA):

- The CTA must be incredibly simple and remove all friction.

  ◇ **Weak CTA**: "We hope you'll consider participating."

  ◇ **Strong CTA**: "To join us in this exciting initiative, please confirm your participation by [Date] by emailing [Name] at [email address] or calling [phone number]. We will follow up with a detailed information packet."

6. The Strategic Rationale ("The Why Behind the Strategy"):

- **Explain how this framework is a long-term investment**:

    ◇ **Resource Multiplication**: Community partners bring expertise, resources, and manpower that schools often lack.

    ◇ **Real-World Learning**: It connects student learning to the world beyond the classroom, increasing engagement and relevance.

    ◇ **Advocacy & Support**: Community members who are actively engaged with the school become its strongest advocates during budget votes or policy changes, creating a powerful base of support.

Final Output Structure: I require the final output to be presented as a ready-to-use professional toolkit.

Output Format:

- **Title**: The Strategic Community Partnership Invitation Framework

- **Header**: Core Philosophy, AIDA Model Explained

- **Section 1**: The Complete Letter Template (Formatted with placeholders [ ] for easy customization, following the AIDA structure)

- **Section 2**: The Personalization & Distribution Protocol (The 3-step guide and multi-channel strategy)

- **Section 3**: Sample Talking Points (For following up with a phone call)

- **Pro-Tips**:

  ◇ **Gratitude**: Always send a personalized thank-you note after the event, ideally written by a student.

  ◇ **Documentation**: Take photos and share them with participants and the wider community to celebrate the partnership.

  ◇ **Sustainability**: Create a simple database of community partners for future initiatives.

Final Request: Please begin by asking me for the Initiative/ Event, Target Audience, Sender's Name/Title, and Desired Community Role. Then, generate the complete Strategic Community Partnership Invitation Framework.

Complete Output

The Strategic Community Partnership Invitation Framework

Core Philosophy: Schools are not islands—they are the hearts of their communities. Inviting community members into our schools fosters mutual growth: students gain real-world knowledge while community members invest in the next generation of leaders.

The AIDA Model Explained:

- Attention – capture interest with a clear subject line and respectful greeting.
- Interest – highlight shared values and why this matters to both school and community.
- Desire – explain the initiative and its mutual benefits.

- Action – provide one clear next step to confirm participation.

Section 1: The Complete Letter Template

Subject Line: Invitation to Shape Future Leaders: Join Maplewood Middle School's Career Day

Salutation: Dear [Community Member/Partner],

Attention & Interest:

We are excited to share that Maplewood Middle School will be hosting our annual Career Day on Thursday, March 14th, from 9:00 AM to 12:00 PM. This event brings together our students and community partners to explore real-world careers, expand horizons, and build connections between classroom learning and future opportunities.ww

Desire (The Ask & Benefits):

We would be honored to have you participate as a guest speaker. By sharing your experiences, insights, and advice, you will:

- Inspire students to dream big and explore diverse career paths.
- Strengthen our community's connection with its schools.
- Showcase your organization's contribution to shaping the future workforce.

Action (The CTA):

To confirm your participation, please RSVP by March 1st by emailing [Coordinator's Name] at [email address] or calling [phone number].

We will follow up with a short information packet that includes logistics, suggested talking points, and student questions in advance.

Closing:

Thank you for considering this opportunity to inspire our students. Together, we can create a meaningful and memorable day for our future leaders.

Warm regards,

[Sender's Name]

Principal, Maplewood Middle School

Section 2: Personalization & Distribution Protocol

3-Step Personalization:

- **Hyper-localize**: Mention the recipient's business, role, or community contribution.
- **Align benefits**: Highlight what they gain (e.g., future talent pipeline, positive visibility, legacy).
- **Authentic voice**: Ensure the tone reflects the school's culture and values.

Multi-Channel Strategy:

- **Primary**: Personalized letter/email.
- **Secondary**: Post event details on school website & social media.
- **Tertiary**: Phone call from a student or PTA member to key partners.

Section 3: Sample Talking Points (for follow-up calls)

- "We'd love to invite you to Career Day as a speaker."
- "Your expertise in [industry] would be so valuable for our students."
- "It's a short commitment (one morning), but it will make a lasting impact."

- "Can I count you in for March 14th?"

Pro-Tips for Sustainability

- **Gratitude**: Send handwritten thank-you notes from students.

- **Documentation**: Take photos and share highlights with participants and the community.

- **Database**: Keep a simple log of engaged partners to invite for future even

Back Matter

Best Practices for Using AI in Education

Adapting Prompts for Grade Level and Subject: Ensure that AI-generated prompts are tailored to the specific grade level and subject matter being taught. This customization helps maintain relevance and accessibility for all students, supporting effective learning outcomes.

Reviewing AI Outputs Prior to Classroom Use: Carefully evaluate all AI-generated content before introducing it in the classroom. Thorough review is necessary to confirm accuracy, appropriateness, and alignment with educational goals.

Encouraging Student-Centered Applications: Promote opportunities for students to interact with AI tools in ways that foster engagement and personalized learning. Student-centered use of AI can support autonomy and creativity in the learning process.

Balancing AI Efficiency with Teacher Creativity: Leverage the efficiency offered by AI while preserving the unique creativity that teachers bring to the classroom. Striking this balance ensures a dynamic and enriching educational experience.

Maintaining Ethical and Responsible AI Use: Uphold ethical standards and responsibility when using AI in education. Prioritize transparency, fairness, and the well-being of students at all times.

Appendix: Prompt Variables Guide

{topic} – Subject or lesson theme

{grade/level} – Grade level or age group

{studentName} – Individual student name

{assignment/project} – Specific task

{timeFrame} – Duration or schedule

{learningStyle} – Visual, auditory, hands-on

{tone} – Formal, friendly, motivational

...and more variables specific to prompts

Further Resources

- **EdTech Tools**: Google Classroom, Kahoot, Quizlet
- **AI Tools**: ChatGPT, Canva for Education, Grammarly
- **Pedagogy Books**: "The Skillful Teacher", "Teach Like a Champion"

Acknowledgments (Personalized Version)

Creating *The AI Prompt Playbook™* has been a deeply personal journey — one shaped by curiosity, late nights, and a belief that creativity and technology can work together to empower people everywhere.

My heartfelt gratitude goes to the educators, creators, and AI enthusiasts whose insights and encouragement inspired the structure and purpose of this book. Your feedback turned early ideas into a guide that can serve both beginners and professionals exploring the world of prompt design.

To my family and friends — thank you for your patience, positivity, and constant support through every stage of this project. Your belief in me has always been the quiet force behind my work.

I am also grateful to the open-source and AI communities, whose innovations made it possible to transform imagination into practical, accessible tools. A special thanks to the teams behind Artificial Intelegince, Adobe Illustrator and Adobe InDesign, whose platforms made this book possible both in content and design.

Finally, to every reader — thank you for bringing *The AI Prompt Playbook™* to life through your curiosity. I hope it inspires you to see AI not just as a tool, but as a creative partner in your own journey.

— Mohammad Taib

Leesburg VA, 2025

About the Author

Mohammad Taib is a forward-thinking educator, creative strategist, and innovator in the field of AI-assisted learning. With a passion for both teaching and technology, he works to bridge the gap between traditional education and the opportunities offered by artificial intelligence.

As the creator of The Wonder Book of Prompts series and the AI Prompt Playbook for Schools & Education Systems, Mohammad is dedicated to empowering teachers, students, and school leaders with practical tools that make learning more engaging, inclusive, and effective.

Born in Afghanistan and currently based in the United States, he brings a global perspective to his work, deeply valuing education as a path toward opportunity and progress. His projects — from building creative prompt libraries to designing AI-powered educational apps — reflect his mission to make technology accessible, supportive, and transformative for all learners.

When not writing or building new ideas, Mohammad enjoys working with youth, exploring creative design, and contributing to projects that promote lifelong learning and personal growth.

The AI Prompt Playbook™

is *dedicated to everyone learning, teaching, and creating their own path with AI.*